Human Aspects of Library Automation

HILARY DYER WITH ANNE MORRIS

Gower

Published by
Gower Publishing Company Limited
Gower House
Aldershot
Hants GU11 3HR
England

Gower Publishing Company
Old Post Road
Brookfield
Vermont 05036
USA

British Library Cataloguing in Publication Data
Dyer, Hilary
Human aspects of library automation.
1. Libraries. Automation. Ergonomic aspects
I. Title II. Morris, Anne, *1950–*
025'.0028

ISBN 0 566 05543 0

Printed in Great Britain by Galliard (Printers) Ltd, Great Yarmouth

Contents

Figures

Tables

Acknowledgements

This book could not have been completed without the help of a number of people: librarians and information officers, ergonomists, photographers, typists, proofreaders, draughtsmen, and general dogsbodies. Our grateful thanks are extended to (in alphabetical order): Stephen Barnacle; Vanessa Corrick; Linda Dorrington; Rick Dowling; Lionel Durbidge; Graham Dyer; Mary Edmunds-Otter; Judith Elwell; Maureen Forrest; Shane Godbolt; Anne Gonsalves; Peter Leggate; Jenny MacDougall; Elizabeth Marsh; Irene Murray; Brian Shackel; Linda Stewart; Sandra Webb; Nicky Whitsed.

Introduction

Automation in one form or another is a dominant theme in the library world of today. The implementation of computer-based systems has received much attention in the literature. Focus has, however, centred primarily on the technical and practical aspects of the process, with relatively little mention being made of the effect on the people who will operate such systems or be served by them. Fine (1982, p. 209) expresses it thus:

> The greatest marvel of technology is that if it breaks down, we can fix it; if it has flaws, we can debug it; if it doesn't work at all, we can ignore it; and if it works well, we can make it work better. No one has as yet figured out a way to debug the human factor. It is the most complicated aspect of any technological system, yet it's the one that gets the least attention, is least discussed, the least researched, and perhaps the least understood.

Many advantages are to be gained by automation, such as increased productivity, improved reader services, faster movement of acquisitions between their delivery and appearance on the shelves, improved stock control and a better quality of working life for all concerned. Nevertheless, there are many pitfalls on the road to successful computerization, which can counterbalance several of these advantages. It is the problems with the human aspect of library automation which this book seeks to address.

The importance of the human factor must not be underestimated. If people respond badly to automation, the anticipated effectiveness of the system will not be achieved. Poor workstation and job design can result in poor health and can induce, or increase, stress. Poorly designed software interfaces result in underused systems and decreased accuracy. Thus, in order to achieve a successful automation project, attention needs to be paid to the human factor.

A vast literature exists on the ergonomics of computers. The term 'ergonomics' is derived from the Greek word *ergon* meaning work and *nomis* meaning natural laws (Oborne, 1985) and is used to describe the study of the relationship between workers and their environment. This book is an attempt to summarize the latest research into the various aspects of human–computer interaction which is applicable to libraries. However, as research is continually being undertaken and books have a long gestation period, it is not possible to ensure that the latest report to hit the headlines will be covered. Every effort has been made to give a thorough examination of the topic, but no guarantee of completeness can be given.

Concentration has been placed on the health and safety factors and workplace design (Parts II to V) because it is believed that these areas are of prime importance. Rumours about the possible risks to health abound in the general and professional press, and it is hoped that the review of the research in this field will clarify some of the issues and allay some of the fears. One of the primary causes of health problems among VDU operators is poor workstation and/or job design. The latter part of the book therefore concentrates on methods of improving the operator's workplace and job design. It is appreciated that lack of funds and inadequate buildings are partly to blame, but it is believed that, if librarians become more aware of the alternatives available, they may be better able to argue for an improvement in these conditions.

The software interface (Part VI) has no small part to play in the overall level of operator satisfaction. With the relatively small choice of library systems available, it is perhaps difficult enough to choose a system which offers appropriate functionality without complicating the issue by assessing the quality of user interface. It is considered important, however, to outline those factors which contribute to or hinder ease of use, if only to heighten the awareness of potential software purchasers and so to stimulate and make more articulate the demand for a more usable interface.

Part VII examines the implications of automation on the operator's work. Changes here may, in turn, affect the design of the workstation to incorporate variations in job procedures. Having studied the detail of what is affected by the automation process, attention is paid to how an automated library system should be implemented. Consideration is given to the management of change, the planning and implementation process and the training of both staff and users.

Although it is recognized that many of the staff involved in library automation at various levels will be female, to ease both reading and writing, however, the term 'he' has been used throughout.

Part I HUMAN CHARACTERISTICS

1 HUMAN CHARACTERISTICS

Introduction

In order to appreciate the effect that computers may have on their users, it is necessary to identify the human characteristics which will be most affected.

The mental and physical make-up of a person affects the way in which he reacts to the world around him. In terms of an automated library system, the surrounding environment will contain several factors which impinge on a person's private world. For example, a change of system will generate stress to a greater or lesser degree depending on the attitude of the person towards the change; the need to learn and remember new procedures will make demands on the memory process; and the need for concentration or attention may alter. Man's psychological make-up is complex and no attempt is made to outline it here. Concentration is placed on aspects of his information processing system, its importance in the automation process and factors which affect it.

Physically, people will also vary. Rather than forcing people to adapt to a given arrangement of furniture, possibly inducing harmful postures, it is desirable to redesign the workplace to take into account the human being as well as the computer. The table at the end of this chapter outlines the main physical characteristics which need to be taken into account.

Man's information processing

People interact with computers by giving and receiving information. The quality of the communication between man and machine will depend upon an accurate exchange of data. For this reason, it is important to consider how man processes information and the factors which affect his performance.

Information is fed into the brain via one or more senses. Once there, it is operated on and then the manipulated data are either stored for future use or transmitted via various output devices (for example, speech or actions).

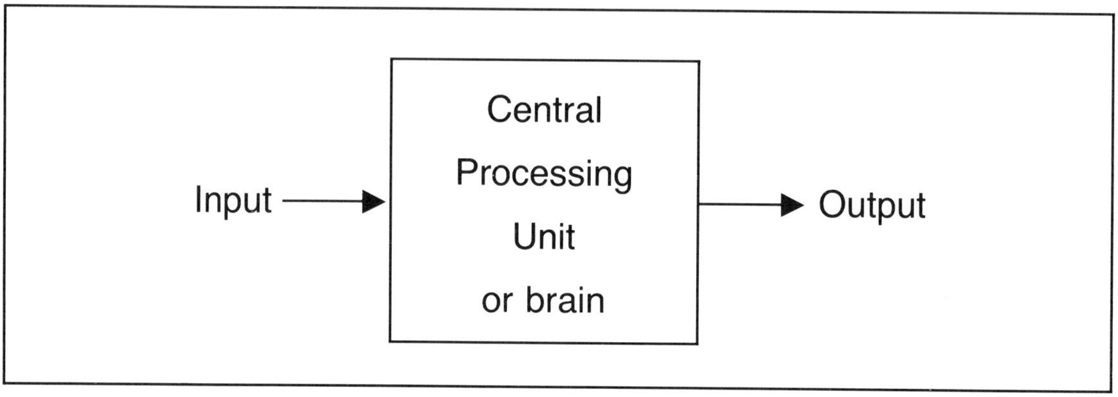

Figure 1.1 Basic information processing model

Data input

Information is received by a person via his many body receptors. These take in data from the outside world, via one or more of the five senses: sound, sight, touch, smell and taste; from inside the body – for example, they inform the brain when the body is tired or hungry or thirsty; and from the neuro-muscular system which gives information to the brain about body position. Once the information has been received by the receptors, it is passed to the brain via the nervous system.

Unfortunately, man's information processing system is not always accurate, and the information can be interpreted by the brain in a distorted manner. Such factors as perception, attention, arousal, motivation, emotion, past experience and environmental features can interfere singularly or interactively with the analysing process and cause the brain to misinterpret the information. Many of the factors affecting man's information processing will be discussed further later in this chapter.

Central processing facilities

The central processing facilities in the brain may be seen as being divided into various mental activities or processes. Information or stimuli are first received and held in a very short-term sensory store. The information then passes to the brain, where it is perceived. The process of perception involves matching the sensory information with a 'template' or similar representation which is held in a long-term memory store. Once the information is identified (assuming it is), a decision has to be made as to what action to take. A response can be evoked immediately and/or the information can be committed to man's working memory. Once in the working memory, information can be simply forgotten, in a relatively short time, committed to long-term memory – that is, learned, or used to generate a response. Responses, either before or after committing information to the working memory, involve some sort of controlled muscular movement as stated above. The consequences of each response are then perceived as feedback and the whole process starts again.

Obviously, this cycle of events takes place rapidly, but it is important to note that there are limits to both the quantity of information that can be processed in a certain time and the amount of 'mental energy' available when undertaking information processing tasks.

Data output

There is very little to say about data output except that it generally takes the form of some sort of controlled muscular movement. This includes not only the movement of limbs but also of the mouth (for speech), face (facial expression) and eyes. Output can be affected by constraints imposed by the working environment, an operator's physical well-being and other factors.

Factors affecting man's information processing

Attention

The concept of attention refers to 'global limitation on the resources available to time-share activities and to process information in parallel' (Wickens, 1987, p.95).

Man is capable of attending to only a limited number of stimuli at any one time. In fact, some theorists propose that man is only capable of attending to one source of information at a time (the single channel theory). Others believe that this theory can be extended. They believe that, although man might only be able to process or attend to one item at a time, he is nevertheless capable of rapidly dividing his attention across several tasks thereby giving the appearance of processing information in parallel. A number of other people contend that man can perform two or more tasks simultaneously, particularly if one of these can be performed without conscious attention in an 'automatic' mode of processing – that is, where stimuli produce a response but are not perceived. However, for managerial and system design purposes it is safer to assume that man can pay attention to only one source of information at a time.

Certain kinds of stimuli or information are more successful in gaining our attention than others. These include stimuli that are of high intensity (a loud noise has the advantage over a low murmur and, similarly, a bright light over a dim one), stimuli that suddenly change (the fan on the computer may go unnoticed until it stops working), stimuli that differ or contrast significantly with their surroundings, stimuli which involve movement (such as flashing error messages) and stimuli which are repeated (the repeated high pitched bleep of an activated library security system, for example, invokes more attention and a better response than if there was only one bleep). Stimuli which have some sort of emotional importance (such as the utterance of a person's own name) also attract attention.

The general level of attention is also determined by internal factors. These include:

- individual habits of attention and inattention (people who have learned to attend to certain inputs and disregard others perform better than those who have not);
- level of interest (the more interesting the stimuli the more likely the person is to respond);
- level of motivation (highly motivated people will attend longer, see 'Motivation' below);
- the perceived relevance or importance of the input (relevant and important stimuli are more likely to be filtered through and attended to than less relevant and less important material);
- the general level of arousal (see 'Arousal' below);
- to some extent, a person's mental capacity (see 'Mental capacity' below).

Obviously, a library manager must be aware of all the factors affecting attention; at one extreme, inattentive staff can be mildly irritating to both managers and clients and, at the other, incur vast unnecessary expense, particularly if unheeded warnings lead to computer breakdowns.

Arousal

People talk about being aroused from sleep – that is, experiencing a change of state from where brain activity is low to one of increased activity or arousal. Research has shown that arousal is correlated to performance levels. Too little arousal (caused by insufficient external and internal stimulation) and too much arousal (excessive external and internal stimulation), for example, cause performance to deteriorate. Optimum performance can only be obtained if the arousal conditions are perfect for a particular operator and task.

External factors include such things as the number of customers using the library (arousal may be very low if there are only few readers or, conversely, too high if there are queues of customers); noise (too much and too little can cause over- and understimulation respectively); and lighting (glare on the VDU screen may overstimulate the eyes). In addition, there are numerous physical, psychological, social and personal internal factors which can raise or lower a person's arousal system and thereby alter performance levels. These include, for example, age, sex, health, personality, intelligence, ability to get on with colleagues, and motivation. The amount of stimulation which should be aimed for will depend on the characteristics of the operator(s) and the nature of the task. There are no data providing optimal arousal/stimulation levels; each case needs to be examined individually. It has been found, for example, that 'piped' music increased the performance of operators doing boring, repetitive industrial tasks (Fox, 1983). It is doubtful, however, that similar success would be obtained in a library! It is, therefore, necessary to examine the circumstances in each library, interviewing the operators and assessing the nature of each task. The aim is to achieve a balance, with neither too much nor too little stimulation.

Motivation

Motivation is the term used by psychologists to describe the study of what induces people to act. An individual's level of motivation affects whether incoming information receives attention and whether it will produce a response. In other words, it will affect people's ability to learn new concepts and to respond to different situations.

Motives have been classified by several different researchers. Maslow (1943), for example, classified motives in relation to a hierarchy of needs. These, in ascending order, are: physiological needs; the need for safety/security; social needs; needs of the ego; and the need for self-actualization. He postulated that man is motivated to satisfy these needs in a bottom-up approach – that is, physiological needs are of paramount importance and must be satisfied before those related to safety and security and so on. Alderfer (1969, 1972) updated Maslow's concept and introduced the following groups of needs:

- 'Existence' or 'basic' needs – related to the survival of the individual.
- 'Relatedness' or 'contact' needs – related to man's social activities.
- 'Growth' or 'self-actualization' needs – related to man's desire to have influence over his situation.

Managers in the Western world have little need to concern themselves with such basic physiological needs as warmth, food and clothing, although they do exercise considerable influence over if, and how, the remaining needs are met, and hence whether or not a person is motivated to carry out his work. For this reason, they need to be sensitive to the factors which affect motivation.

Herzberg, Mausner and Snyderman (1959), while investigating work incentives and the working environment, found that the most important 'satisfiers' or 'motivators' were related to performance, recognition, work content, responsibility, promotion, and pay. Others included organizational policies, relations with subordinates, colleagues and supervisors, and working conditions.

In general terms, people like to feel that they have done a good job, in order to have a sense of achievement. This can only be attained if they have met goals either set by themselves or demanded by the task or organization. Recognition of effort, or feedback, is extremely important. Sometimes a 'pat on the back' is all that is necessary. In other cases, recognition might be achieved by promotion. Work content involves not only what the job entails, but how rewarding it is, the degree of variety, the level of complexity, how much stress it causes, and the degrees of control and autonomy it provides. All these need to be considered when undertaking job design and are discussed more fully in Chapter 19. The level of responsibility can be a motivator or demotivator, depending on the individual. Responsibility to one person may cause stress and act as a demotivator, but to another may provide an opportunity for self-fulfilment. Consequently, managers must be cautious when using this approach as an incentive or motivator. Most people like to feel that there is some chance, at least, of promotion. Not many people are content with 'dead-end jobs', hence some sort of career structure within organizations is welcomed by most employees. Low pay can drastically reduce motivation and thus the performance of an individual. Unfortunately, low pay is one of the characteristics of much library and information work. Fortunately, in many cases, this demotivating factor may be counterbalanced by the enjoyment many professionals obtain from their work.

Organizational issues, such as how the organization is structured and whether management are caring, accessible and approachable also affect an employee's motivation. Similarly, relations with subordinates, colleagues and supervisors play a part. Poor relations create a bad atmosphere and result in poor motivation. Working conditions – such as the number of hours worked, the amount of holidays, restaurant facilities and the working environment – also affect the degree of motivation. Obviously a safe, spacious, pleasant, noise-free environment is more motivating than one that is not. The last factor mentioned by Herzberg *et al.* (1959) was that of leadership. It is a fact that most employees prefer a democratic rather than an autocratic form of leadership. People, on the whole, like to feel they are involved in an organization and that they can contribute to its overall success.

It can be seen, therefore, that an individual's level of motivation depends on many factors, their relative importance differing for different people. For some, it is sufficient to have good pay and conditions; for others, an interesting job may be more important. Motivating factors also interact and counterbalance each other. For example, a librarian may have to work in extremely cramped conditions, be poorly paid and overworked, yet still be fairly motivated because he enjoys his job, thinks it is rewarding and gets on well with his colleagues. When assessing whether staff are motivated, managers must consider, therefore, many factors and determine the overall combined effect. If motivation is low, then ways of increasing it must be sought, such as providing encouraging feedback, improving job design and the work environment or creating more channels for self-enhancement. It must be remembered that increasing the pay of employees can be a useful and welcome motivator in the short-term, especially when pay is low, but is rarely satisfactory, on its own, as a long-term solution to poor motivation.

Mental capacity

The mental capacity of individuals varies enormously, but there is one common factor: namely, that if you exceed a person's mental capacity, performance deteriorates rapidly. Brown and Poulton (1961) and Brown (1965) have expanded this fact into a 'mental capacity and performance' model. Essentially, they suggest that performance depends on a trade-off between the demands of a task and of the environment, and the operator's own abilities. They believe that a person has a limited capacity to perform tasks. That capacity may be consumed by the demands of the job or by operator effort in compensating for an unpleasant environment or poor equipment and thus performance will drop. Thus, in order to maximize an operator's performance, it is necessary to match the demands of the task and the suitability of the environment to the operator.

Perception

Perceptual recognition can be described as the association between an incoming set of information or structure, and a recognized 'template' that is stored in long-term memory (Neisser, 1967; Rummelhart, 1977; Wickens, 1987). For example, people can perceive written words because they have a template of what each word means in their long-term memories. People's perception is affected by various factors, with the result that one person will perceive things differently from another.

Perception is affected by the overall quality of the stimuli presented and a person's general receptiveness at that particular time. The quality of the stimuli is affected by a large number of factors, such as the duration and frequency of their presentation. In the case of written cues, the size of the text, glare, display characteristics, character representation and the format (VDU, book and so on) also have an effect, as do the duration of an event, its loudness, intonation and accent with verbal stimuli. A person's general receptiveness is equally affected by many factors – for example, age, motivation, level of arousal, intelligence, attentiveness, health, and his familiarity with the stimuli. In addition, environmental factors such as noise level, the presence of vibration, inadequate lighting and poor thermal conditions can affect both the quality of the stimuli and a person's overall receptiveness.

In order to ensure that performance is not degraded, it is necessary to increase the likelihood of incoming stimuli being perceived. Thus, it is important to ensure that the stimuli presented to the person are not distorted in any way and that conditions are such as to enable the person to remain alert and receptive. In other words, managers must consider carefully the design of software, hardware, workstations, jobs and the environment to ensure optimal conditions. All these are discussed in considerable detail in later chapters. Perception can also be increased, to some degree, by increasing stimulus expectancy (Wickens, 1987). For example, an online searcher can often read the content of a line of information even though it appears only momentarily on the screen. This is because he expects to receive certain prompts and can recognize them at a glance by the overall shape and pattern of characters. Stimulus expectancy, in this case, is increased by the familiarity of the material.

Perception of incoming stimuli is not just an automatic process; it involves much brain activity and its success depends on many factors of which managers need to be aware. Possibly the most vital of these is that people will perceive things differently.

Memory

Memory is an abstract and ambiguous word which, 'in its most comprehensive sense, refers to the effects of a person's past on his present' (Hunter, 1964, p.19). A person is modified by what he does, and these modifications, if retained, will affect his behaviour sometime in the future. Consequently, modifications or past experiences that are held in the memory enable people to adjust to new circumstances. Psychologists usually think of memory as having two components: short-term memory (or, preferably, working memory) and long-term memory. Working memory is regarded as a temporary and fragile store where we keep our limited 'conscious' information, whereas long-term memory is believed to be where we store more permanent facts and experiences. This section concentrates on short-term memory, long-term memory being related very closely to learning which is discussed in depth under 'Learning'. The main characteristics of short-term memory are as follows:

- The length of time information can be held in working memory without repetition or rehearsal is extremely short; approximately 10–20 seconds (Peterson and Peterson, 1959).
- Information may be maintained for longer if rehearsed, but competition from other perceptual and cognitive activities may cause attention to waiver (Klatzky, 1980; Underwood, 1976).
- The number of unrelated items that can be held even when full attention is given is limited to between two and seven. A larger number than this may well exceed capacity and result in items being forgotten or transposed (Miller, 1956).
- Even a small number of items will be forgotten within 20 seconds if this interval is filled with some irrelevant, distracting activity (Deese, 1964).
- A technique called 'chunking' may be used to increase the amount of information stored. Chunking involves grouping individual items. It has been found that related 'chunks' of letters or digits can count as single items. For example, words such as 'main exit' may be counted as two items and the area code of a telephone number as one item even though it may contain more than one digit.
- Meaningful information is stored more successfully than meaningless information. Strings of characters, for example, are more readily remembered than strings of numbers (Conrad, 1964).
- Long strings of unrelated digits are more easily remembered if there is a gap after every third or fourth digit. For example, 333 6195 will be more easily remembered than 3336195 (Wickielgren, 1964).
- Retention is better if related items are grouped separately from unrelated items; for example, telephone numbers are frequently divided, the area code being separated from the rest of the digits.
- Similar items are more likely to be forgotten than dissimilar items. For example, the acoustically similar string EGBZCDT is less likely to be recalled accurately than XFEYWMU (Conrad, 1964), and letters or words that rhyme are particularly difficult to remember.

- A memory load of combined spatial (pictorial/visual properties of the environment) and verbal (linguistic properties) items will be better retained than the same number of either spatial or verbal items exclusively (Kyllonen and Alluisi, 1987).
- Memory for verbal material will be disrupted more by ongoing perceptual, cognitive and other activities that are also verbal and vocal than by simultaneous spatial activities. For example, it is difficult to listen to the radio and hold a telephone conversation – both verbal activities – but relatively easy to knit whilst listening because knitting is a spatial activity. The converse is true for spatial activities (Baddeley and Hitch, 1974; Baddeley and Lieberman, 1980; Wickens, 1984).
- Working memory can be improved using the technique of visual imag-ery where a set of items is associated with distinct visual images which are already held in long term memory (Bower and Reitman, 1972).
- The older the user – particularly if over the age of 55 – the less information he can hold in working memory at any given time (Deese, 1964).

Some implications for designers and managers

System designers and managers must be aware of the limitations of human working memory and not overload it.

- Where possible, people should not be expected to memorize more than seven unrelated transient items. In computerized menus, for example, the number of options offered should ideally be equal to, or less than, seven but failing this should never exceed nine. It is appreciated, however, that when a large number of choices are to be made, this can result in a frustrating number of menus.
- To facilitate 'chunking', instructions should always present related information together.
- Accession numbers and other long numbers such as class numbers which need to be remembered should have gaps after every third or fourth character. This is frequently the case when the Dewey Decimal System is used when the decimal point is placed after the first three digits, but close classification often results in long strings of numbers following the point.
- Where heavy demands are made on working memory, distractions should be kept to a minimum.
- If long codes, such as passwords or logon names, have to be memorized they should be meaningful. If they are not memorable, the operator may be tempted to write them down and display them near the terminal, which somewhat defeats the objective of security codes.
- Important messages should be both verbal and spatial (pictorial). For example, handler warnings on floppy disks frequently contain pictures as well as written instructions.

- Allowances must be made for employees over the age of 55 (or even younger!).

Learning

Learning can be defined as 'a process which optimizes performance based on the recognition and use of familiar features in a present situation' (Gibson, 1965). It involves recognizing patterns in a situation, generalizing from one situation to another and making selective responses that optimize performance. The first stage is essentially an exercise in matching selected features in a current task with features experienced in the past. Gibson (1965) suggests two ways this might be done. First, by template matching; features of the current situation are compared and matched with previously learned 'templates'. Second, by property listing; this involves the elements or features of the present situation being compared with stored lists of elements and features. In practice, a combination of the two methods might be used. However, whatever the method used in the recognition stage, it is extremely unlikely that a perfect match can be made. Consequently, learning must involve the processes concerned with generalization and selection of appropriate responses, which are linked to man's decision-making mechanisms (see the following section).

Given the same instruction, some students or trainees learn more quickly than others. In part, the difference can be attributed to age, general aptitude and past experience. However, it is a fact that some people have better learning skills and can assimilate new information more quickly than others with similar age, aptitude and experience characteristics. By examining these learning skills or learner strategies, a library manager can assist staff to improve their learning capacities. This could be particularly important if people will need to operate new systems or adopt different working practices.

There are many cognitive psychology textbooks as well as study guides which detail learner strategy principles, for example: Anderson (1980), Gage and Berliner (1979) and Mace (1968) to name but a few. Kyllonen and Alluisi (1987) also provide a useful overview. The following is a summary of these principles and factors which affect man's ability to learn.

- *Meaningfulness of information*. Meaningful information is much easier to assimilate than meaningless data. An understanding of the overall system will facilitate the learning of smaller modules. Similarly, the digestion of a manual's overview will help users locate items of interest more quickly.
- *Memory load*. There are severe limitations on the amount of information man can process at any one time (see 'Attention', p.5 and 'Memory', p.10). A person's working memory quickly becomes overloaded when bombarded with incoming stimuli. Receiving new information is essential to learning. It is therefore important to optimize the quantity of information received to maximize learning

but to ensure that the memory is not overloaded. Fortunately, strategies exist which can reduce memory burden. Two of the most important are rehearsal and chunking. Rehearsal involves simple repetition of the incoming stimuli, such as repeating a telephone number or the name of an enquirer, whereas chunking is concerned with organizing incoming data into familiar and meaningful groups or chunks. Other important ways to reduce the processing overload are to make some of the basic skills required automatic by practising them over and over again, and the use of external memory aids such as documentation or online help.

- *Practice (or reinforcement)*. The old adage 'practice makes perfect' has been verified in numerous experiments. All show that the more the material is studied, the stronger will be the memory trace for that material (Wickielgren, 1981).
- *Recitation*. Reciting the main points at the end of a training session or a chapter in a textbook aids learning (Anderson, 1980).
- *Spacing of practice sessions*. Evidence suggests that learning is improved if practice or training sessions are spaced over a period of time rather than in one long block. For example, 10 one- or two-hour sessions spaced over a week will produce better results than 10 or 20 hours consecutively (Mace, 1968). Adequate rest breaks are essential. Research has shown that concentration wavers after a short period of time. Taught instruction, for example, should last for no longer than 40 minutes at any one time (Mace, 1968).
- *Mix of topics*. An effective learning strategy involves a good mix of topics – for example, one or two sessions on one topic, two sessions on the next and so on. Concentration on one topic reduces learning capabilities.
- *Parts-learning*. Often learners try to 'run before they can walk'. It is much more efficient to master one or two skills at a time before going on to the next.
- *Expert performance*. One way of facilitating the learning process is to observe an expert and then to analyse and copy his performance. Obviously, this method is not appropriate in all situations but 'sitting by Nellie' techniques have been used successfully for decades.
- *Feedback*. Learning is increased if feedback is given. The feedback may be negative, indicating that an error was made, or positive in the case of correct response. Positive feedback is to be preferred as it serves to strengthen and reinforce a learner's response. The sooner the feedback is given, the more effective it will be. (See also 'Response mechanisms', p.16.)
- *Extraneous distractions*. For effective learning to take place, the learner must commit the material (verbal, written or actions) to long-term memory. Anything which interrupts this process – for example, interactions with staff or customers or the need to answer the telephone – will have an adverse effect.
- *Stress*. The overall anxiety state or mood of a person can affect how

much is learned; the more stressed a person is the less he learns. The use of relaxation exercises and positive thinking techniques can greatly enhance performance (Meichenbaum, 1977; Velton, 1968).

- *Motivation*. Whether a person is motivated to learn can affect whether incoming information is attended to. Hence motivated people are likely to be more receptive to new data. One way to motivate staff is to make the task interesting. Malone and Lepper (1983) suggest that instructional programs, for example, can be made more interesting if they incorporate some of the characteristics of computer arcade games, which have proved to be extremely motivating. These characteristics include the availability of short-term goals, the use of immediate feedback (scores), the adaptability of the task to the user's skills level, and other characteristics such as exciting graphics and animation. Other factors will include the degree of relevance the trainee sees between the task to be learned and their daily work as they perceive it.
- *Transfer of training*. The transfer of training is the term used by psychologists to describe the influence of past learning on present learning. If the influence is positive, then positive transfer of training is said to have taken place and the rate of learning is accelerated. On the other hand, if past learning interferes with the learning of new material then negative transfer of training occurs and learning suffers. An example of the latter is when the commands of one system do not mean the same on another, such as is the case between different online hosts and databases, or when a new library system is installed which utilizes conflicting commands to the previous system. Thus, extra time needs to be allowed if there are elements of negative training or, if possible, negative influences should be eliminated.
- *Active versus passive learning*. Active learning is more effective than passive learning. People learn more far more quickly if they become involved in the learning process – for example if they write notes or ask questions during a training session. Passive observation (sitting back and expecting all the information simply to be absorbed) can increase the probability of boredom and of inattention (Schneider, 1982, 1984). Thus, where possible, staff should be actively involved during training sessions.
- *Training context*. The learning process is enhanced if training occurs on-site rather than elsewhere. One of the reasons for this is that the environment provides memory cues for the material learnt. However, it must be remembered that distractions and interruptions reduce learning and if these are likely to occur, as they would on a library desk during opening hours for instance, consideration needs to be given to an alternative time or place.
- *Pictorial/graphical information*. Pictorial and graphical information is learnt more quickly than written or verbal information. For effective learning to take place it is advisable to have a mixture of different

types of information presented in a variety of ways – for example, overhead slides, films, tutorial sessions and so on.

Some of these characteristics will seem obvious, but it is surprising how easily they can be forgotten when it comes to designing training sessions. Careful planning can result in training itineraries which not only provide more effective learning but happier and more motivated staff. This is discussed again in Chapter 19.

Decision-making

Decision-making is formally defined as 'the selection of an alternative response to an inferred environmental situation about which complete certainty is lacking' (Schrenk, 1969). Typically, the types of tasks which involve decision-making are those that demand high-level intellectual skills, including setting objectives, allocating resources, diagnosing environmental conditions, defining and choosing courses of action and interpreting patterns of events. The role of decision-making in the automation process is twofold. First, a decision needs to be made as to the best system to buy and, second, an appropriate automated library system can provide better data upon which to make informed decisions about library management.

The need to understand more about man's decision-making processes is therefore important. The role of the decision-maker has become complicated by the vast increase in the amount and variety of information that has become available through the new capabilities for gathering, processing and displaying data. Time allowed for making decisions has also been reduced as both workload and communication capabilities have increased. Instant decisions are often the order of the day. As technology and system complexity have increased so, too, have the penalties for making a wrong decision, both from a financial standpoint and in terms of the health and safety of fellow employees. Consequently, it is extremely important for any manager to be aware of man's limitations as a processor of information, in order to make allowances for these in system design. Equally important is the knowledge about how decision performance can be improved.

It is well known that man can assimilate and process information only at relatively low rates and is poor at remembering information and doing computations. What is not often taken into account are the following characteristics and limitations:

- There are significant differences in decision-making skills among people; those with 'open' minds are better than those with 'closed' minds.
- People tend to want too much, rather than too little, information to assist them with the result that they then 'cannot see the wood for the trees'. This is particularly true with management information from automated library housekeeping systems which fail to provide information tailored to a library's requirements.

- Too much weight is given to information collected early in the process; this is called the 'primary' effect. This affects aspects of system design, such as the menu, where options should be displayed in decreasing order of likely selection.
- Opinions tend not to be revised on the basis of new data; people are conservative in nature.
- People are unable to make full use of available information, particularly multi-dimensional information.
- Decision-makers often consider too few courses of action; people generally opt for 'least effort'.
- Man is not very good at estimating the probability of some types of outcome. He tends to overestimate low probabilities (for example, 0.01 for 0.0001) and underestimate high probabilities (for example, 0.9 for 0.9999), for instance. In addition, he tends to comply with the gambler's fallacy syndrome, which is that an event that has not occurred recently becomes more likely to occur. For example, if a couple have six daughters then many people would predict that their next child must surely be a son. This is not true, however, since each birth is an independent event.

Despite these limitations, people seem to perform surprisingly well at decision-making. It appears that man has an implicit understanding of expressed probabilities and can specify outcome probabilities with reasonable accuracy, particularly in the middle ranges. He can usefully apply probability statements for decisions regarding non-frequent events and can judge the potential value of new information. Furthermore, it has been found that people usually give the greatest weight to the most important factors in decision-making problems. Overall, this means that people can give consistent and reasonably accurate results when undertaking some (most notably, simple) decision tasks. However, man's performance deteriorates rapidly when he has to evaluate many different alternatives and outcomes. Consequently, managers must recognize the possible need to improve the performance of the human decision element in a system.

Response mechanisms

It was noted earlier that memory affects the way in which we respond to events in the present. Thus the memory of previous feedback (see also 'Learning', p.12) will affect our actions in the present.

Feedback during a task is extremely important. Without it, responses can become inaccurate and motivation can be reduced. Feedback can be specific, like the number of 'hits' when searching online or general, for example, a few words of encouragement. Although positive feedback should be encouraged, even negative feedback is better than no feedback at all (Mace, 1968).

VDU users often find that not enough 'help' information has been provided to enable them to make a correct response. Adequate feedback in such cases would not only improve performance but make the task much less stressful and irritating. The total time necessary to respond to a stimulus (reaction time) depends on the length of time necessary to recognize a signal, to make a decision and to make an appropriate movement. It will vary not only between individuals but with how aroused, motivated and attentive a person is at a particular time, and with the type of sensor receiving the stimulus. The average time for reacting to a spoken message, for example, is 150 milliseconds compared with 200 milliseconds for the same visual message. People can respond quicker than this if both sensors are alerted at the same time. Designers need information about reaction times in order to optimize emergency procedures and system-paced tasks (as opposed to self-paced tasks) so that stress is avoided. Reaction times also have important implications for software design (see Chapter 15).

While man's response mechanism is affected by all the characteristics discussed so far in this chapter, it is nevertheless important to emphasize the effects of fatigue and boredom. Fatigue is defined as 'a reduction in a muscle group's ability to contract' (Van Cott and Warrick, 1972). Symptoms of the onset of fatigue are not only physiological in nature; they include tension, stress, tremor, eye strain and, in some cases, pain. Fortunately, much fatigue can be reduced and sometimes even eliminated by redesigning the job, modifying the environment or simply changing the number and spacing of rest breaks. These are extremely important in VDU work and, for this reason, are considered in more detail in Part IV and in Chapter 19. Boredom, like fatigue, may adversely affect performance. Although it is difficult to measure (only subjective measurements are possible), it is known that well designed tasks, software and equipment help to postpone the onset of boredom and reduce its severity and duration. In the case of library and information work, for example, tasks which some may consider boring and repetitive, such as issuing books, may be made more interesting if interspersed with other tasks like answering enquiries.

In summary, the speed and accuracy of a person's response mechanism depends on many individual and general factors. Improvements can be obtained if individuals are given adequate feedback to responses; auditory and visual senses are altered at the same time – especially applicable in emergencies – and fatigue and boredom are kept to a minimum.

Man versus machine

Many of the previous sections in this chapter have focused on man's characteristics and, in particular, his limitations. This section concentrates on the characteristics and limitations of the computer and how jobs should be allocated to humans or to computers in order to optimize performance.

Many authors have compiled lists comparing human capabilities with machines (for example, Chapanis, 1960; Murrell, 1971; and Fitts as referenced in Shackel, 1984). Summarizing these, it appears that man:

- is better at making complex decisions, particularly when unexpected events occur;
- is capable of improvising;
- has a high tolerance for ambiguity, uncertainty and vagueness;
- can rely on a wealth of past experience;
- is able to interpret complicated visual material which involves depth, space and pattern;
- is able to communicate in a variety of ways (touching, seeing, and through bursts of emotion);
- is adaptable;
- is able to maintain himself.

Computers, on the other hand:

- can process vast amounts of data quickly;
- are able to deal with predictable events in a reliable and efficient manner;
- are useful in hazardous environments;
- have infinite patience;
- do not have temperamental moods or social problems;
- can store and access vast quantities of information accurately and quickly;
- can maintain a steady performance – that is, they are not distracted or affected by fatigue or boredom.

In short, computers are eminently suitable for routine tasks that involve the accurate storing and retrieving of vast amounts of information such as the names and addresses of borrowers, catalogues of books in stock, and so on.

When automating a library or information system, however, consideration should be given to the affect on the human operator, the computer being used only to serve and extend his capabilities (Shackel, 1984). As such, it may be unwise for management to attempt to automate *all* possible tasks without due consideration of the effect it will have. For example, some tasks may be undertaken equally well by both people and computers. In these cases, managers should consider not only the cost-effectiveness of the task but the effect on the job. It may not be cost-effective to automate a particular combination of tasks if the result is to reduce staff morale and consequent productivity. The correct allocation of tasks to machines and humans is essential to people's general well-being and overall performance.

Physical characteristics

People vary not only in their mental characteristics, but also in their size and shape and their ability to lift and to reach. Obviously, these differences must be considered when designing equipment, environments and tasks.

The study of the structural differences in man is called anthropometry. Table 1.1 lists the bodily dimensions which are most useful in designing a workstation. It shows the variation in measurement between men and women and between different racial groups. It should be noted that data for Hong Kong Chinese and Indian people show a smaller stature than for the average British person. Unfortunately no figures were given for people from the Afro-Caribbean community who, generally speaking, are taller and longer-limbed than their British counterparts. Libraries with staff and users from these racial groups may need to allow for the differences in size when designing workstations. Disabled people and children also have special requirements which need to be taken into account.

The data in the table show the static measurements for the 5th and 95th percentile of the population. If a randomly selected group of people were measured and the figures plotted on a graph, the following shape (normal distribution) would probably appear (Figure 1.2).

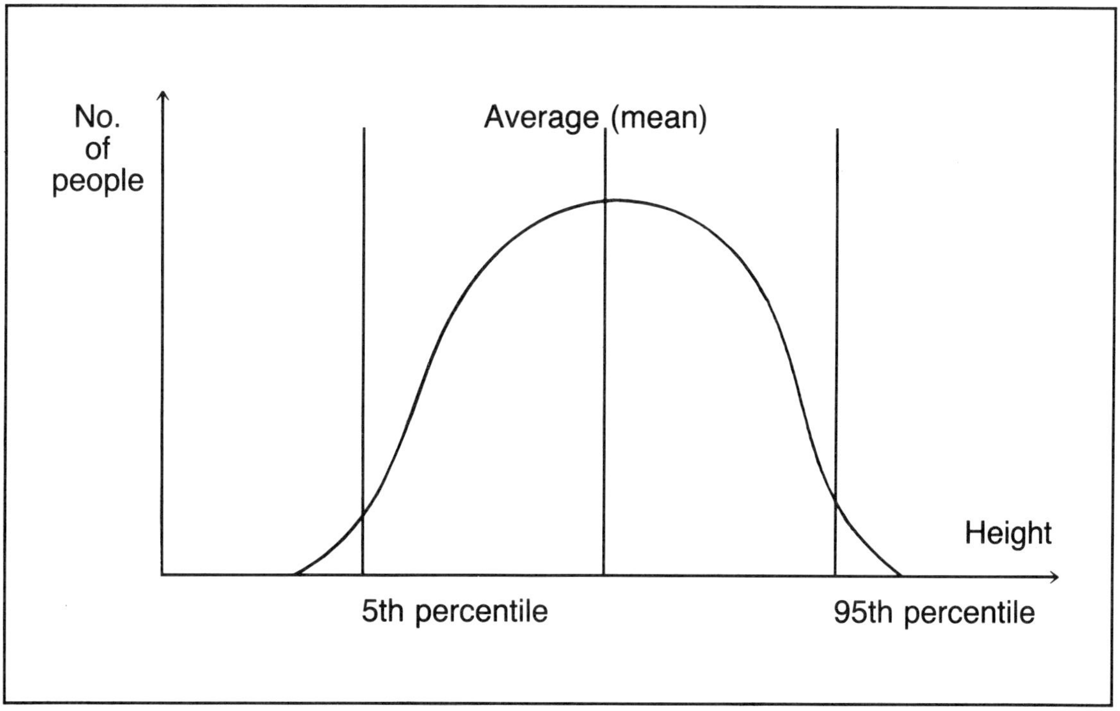

Figure 1.2 Normal distribution

The majority would fall into the middle range – that is, an 'average' person. At the top end of the measurement scale, the number of people registering that height (say greater than 1710 mm) would be very small. Similarly at the bottom end of the measurement scale, few people would

Table 1.1 Anthropometric estimates for various categories of library staff and users. Figures, in mm, indicate the 5th and 95th percentile. (Adapted from Pheasant, 1986).

Part of the body	*Application*	*Correction factor*
Stature	Basic reference	Shoes. Add ~25mm for males and ~45mm for females
Eye height	Standing posture visual height	Shoes. Add ~25mm for males and ~45mm for females
Elbow height	Determines work surface heights	Shoes. Add ~25mm for males and ~45mm for females
Fingertip height	Lowest acceptable level for finger operated controls	Shoes. Add ~25mm for males and ~45mm for females
Sitting eye height (from seat surface)	Sitting posture visual height	Add 10mm for heavy outdoor clothing, reduce by 40mm for 'sitting slump' and seat compression
Sitting elbow height (from seat surface)	Height of arm rests; determines work surface height	Add 10mm for heavy outdoor clothing, reduce by 40mm for 'sitting slump' and seat compression
Thigh thickness	Clearance between seat and the underside of tables	Add 35mm for heavy outdoor clothing
Buttock – front of knee length	Clearance between seat back and obstacles in front of knee	Add 20mm for heavy outdoor clothing
Buttock – back of knee length	Defines maximum acceptable seat depth	Add 20mm for heavy outdoor clothing
Knee height	Clearance required beneath the underside of tables	Shoes. Add 25mm for males and 45mm for females
Floor to back of knee height	Defines maximum acceptable seat height	Shoes. Add 25mm for males and 45mm for females
Hip breadth	Width of seat should not be much less than this	Add 10mm for light clothing; 25mm for medium clothing; 50mm for heavy outdoor clothing

British Adults (19–65yrs)		British Elderly People		Hong Kong Chinese		Indian	
Men	*Women*	*Men*	*Women*	*Men*	*Women*	*Men*	*Women*
1625–1855	1505–1710	1515–1765	1400–1630	1585–1775	1455–1655	1535–1745	1415–1615
1515–1745	1405–1610	1410–1660	1305–1535	1470–1640	1330–1520	1440–1630	1320–1520
1005–1180	930–1085	935–1120	860–1030	950–1080	870–1000	950–1110	875–1025
590–720	560–685	550–690	515–660	575–705	540–680	560–680	525–655
735–845	685–795	675–805	610–755	720–840	660–780	685–795	635–745
195–295	185–280	160–270	150–255	190–290	165–295	180–280	175–265
135–185	125–180	120–175	105–170	110–160	105–155	85–135	80–135
540–645	520–620	510–620	490–600	505–595	470–570	510–610	490–590
440–550	435–530	410–530	405–515	405–495	385–485	415–515	405–505
490–595	455–540	455–570	430–530	450–540	410–500	465–565	425–515
395–490	355–445	365–470	330–430	365–445	325–425	370–460	330–420
310–405	310–435	290–395	285–425	300–370	295–365	275–365	270–390

be recorded. Rather than include the very tall and very short people in the figures, which could complicate matters, a point a little way up on the measurement scale below which lie only 5 per cent of the population and a point near the top of the scale below which lie 95 per cent of the population are chosen. These are known as the 5th and 95th percentiles respectively.

Functional measurements are those taken when people assume common movement positions, such as reaching out in a horizontal plane, reaching as high or as low as possible, or bending over. Since much library work involves stretching and reaching of some kind – for instance when issuing and discharging books and when cataloguing and so on – consideration of functional measurements is extremely important.

Part II HEALTH AND SAFETY CHARACTERISTICS

2 THE IMPORTANCE OF HEALTH AND SAFETY

Why health and safety matters

As automation of libraries and information centres becomes more common, the need to monitor the health and safety aspects of VDUs becomes more important. Two surveys were conducted by the Union of White-Collar Workers, Apex (Apex: London and Home Counties Area Technology Sub-Committee, 1985). The union studied companies in which it had members in both 1981 and again in 1983–84. These surveys examined, among other things, health-related complaints in connection with new technology. 'The result of these two surveys confirmed our worst fears,' the report reads, 'namely that the number of complaints were on the increase and little or nothing was being done to resolve them.' (p.1)

In addition, reports of adverse health effects resulting from VDU work are being reported in the library literature (Steele, 1986), and will continue to be so unless librarians become aware of the causal factors and seek to address them. It was noticeable, in some of the libraries visited during the preparation of this book, that there was a lack of awareness of health and safety factors.

There are three good reasons why the library manager should be concerned with the health effects of automation on his staff. The first is that he has a legal obligation to take 'reasonable care for the safety of his/her employees under the health and safety at work etc. Act 1974' (Suter, 1985). The second and third reasons are suggested by Edward Cluff (1985, p. 30):

> [the second] is that we . . . should care about others. [The third] is that it is irretrievably bound up with productivity. So if we should be insensitive enough not to care about others, we should at least care about matters which affect productivity. If staff feel tired or stressed, whether real or imagined, it will always affect their productivity. They may even have days off from work.

Whatever the reason for wishing to reduce adverse health effects, some understanding of the problems which have been reported and the arguments surrounding them will be helpful, if only to allay staff fears.

Research into these factors is continual and new discoveries are constantly being made. In order to obtain a complete and up-to-date picture, therefore, it will be necessary to scan the relevant journals. The following, it is hoped, will provide a basic explanation as well as the known position at the time of going to press.

Legal aspects

Librarians have a legal duty towards their staff. Suter (1985) reminds us that the employer is both 'under a duty to take reasonable care for the safety of his employees' at common law under the law of negligence and 'to ensure that reasonable care is taken for the health, safety and welfare of his employees at work' in accordance with the Health and Safety at Work Act 1974. The difference between the two is that duty under the Act carries criminal law sanctions, whereas the common law of negligence is a civil matter. Thus, work which causes long-term damage may result in a claim from the employee under civil law. Even symptoms which are not long-term, however, may be brought to the attention of the Health and Safety Executive should employees feel sufficiently strongly. In such a case, the Executive may insist on improvements to the work environment or, in extreme cases, prosecute.

At the heart of both these laws is the word 'reasonable'. This means, in other words, that the risk and degree of damage which might occur and the cost of minimizing that likelihood have to be balanced. In addition, it is important to consider the existing state of knowledge at any time. If risks are not known, then it is unrealistic to expect employers to take steps to prevent them, but where the likely effects are well documented, and the cost of prevention is low, it is reasonable to expect the employer to take preventive measures.

The alleged health hazards of VDU work

The following chapters discuss the hazards to health purported to be caused by VDUs. Although not mutually exclusive, for ease of discussion, they have been grouped into the following categories:

- *Postural effects,* such as backaches and other disorders caused primarily by incorrect posture or workstation design (Chapter 3).
- *Eye and vision problems,* including tired eyes and difficulty in focusing (Chapter 4).
- *Stress-related problems,* such as high blood pressure or irritability caused by the change in the job or fear of automation (Chapter 5).
- *Reproductive hazards.* Reported 'clusters' of miscarriages and their alleged and possible causes (Chapter 6).
- *Other alleged health hazards* including cataracts, facial rashes, photosensitive epilepsy, Legionnaire's Disease, cabling, fire, laser printers (Chapter 7).

3 POSTURAL PROBLEMS

Introduction

A survey carried out by *Health and Safety at Work* magazine quoted by MacMorrow (1987) found that of the VDU workers studied, 53 per cent complained of painful neck or shoulders, 43 per cent had back pain of some kind, 14 per cent experienced hand cramps and 58 per cent suffered from general fatigue. Although this survey and others like it have been condemned as unscientific and biased ['Irresponsible reporting derides VDU operators' health problems', 1986], they serve to indicate that there is a fear, substantiated or not, of physical problems arising from VDU work. Other symptoms which have been reported in the literature include: headaches; pains in the joints, especially in the knees; pains in the legs and ankles; circulatory problems, producing a loss of sensation in the extremities; varicose veins; and bowel and bladder problems.

Not all the physical problems can be attributed directly to VDU work, but it is undoubtedly true that some can. These are problems which are caused by incorrect loading of the muscles as a result of poor workstation and job design, and poor posture which causes restricted circulation. The causes may be grouped as follows:

- forward leaning posture
- unnecessary head and neck movement
- constrained posture causing poor circulation and fatigue
  intensity of work

Forward leaning posture

The operator may lean forward for a variety of reasons, such as to view the screen or source document, to compensate for poor vision or for a poorly designed workstation, or to eliminate glare from the screen. As the head is supported with least muscular effort only when it is vertical – that is, when the cervical vertebrae are erect (Cakir, Hart and Stewart, 1980), a forward leaning posture places a strain on the muscles in the neck and shoulders. This posture also compresses the abdomen and lungs, reducing the efficiency of the digestive system and causing shallow breathing which reduces the supply of oxygen to the body.

Unnecessary head and neck movement

Incorrect positioning of the screen and source document can cause unnecessary, repetitive movement of the head and neck, thus tiring the neck and shoulder muscles, especially if this occurs in a forward leaning position.

Constrained posture

A constrained posture, especially over long periods, can also result in pain in the lower part of the body, particularly in the knees. It may also impair the flow of blood, possibly resulting in a loss of sensation in various parts of the body. Good circulation is necessary to remove waste products from the muscles and thereby to reduce fatigue. Circulation can be improved by minor postural adjustments, such as stretching and changing position, but the operator may not be aware that this is necessary and, when under extreme pressure or concentration, may be less likely to fidget. Disabled people, too, may find this difficult.

A poorly designed chair can compound the problem of a constrained posture by causing poor weight distribution on the thighs. This restricts the blood supply and may cause pooling of the blood leading to varicose veins.

Intensity of work

Hünting (1984) studied four groups of bank employees: 53 data entry operators; 109 conversational terminal operators; 78 typists (copy and audio); and 55 traditional office workers. He found varying degrees of bodily and eye pains in each group as shown in Figure 3.1.

The highest incidence was in the data-entry group and the lowest in the group of traditional office workers. The noticeable contrast between the two groups was that data-entry operators worked intensely at high speeds and traditional office work involved a great diversity of movement. Hünting concluded that the constrained posture, encouraged by intensive keyboarding (Hünting's subjects used only the numeric key pad, with the other hand in a fixed position, turning over source documents), was a causal factor of the symptoms. MacKay (1985), referring to Hünting's work, suggested that both the nature and the duration of the task are to blame.

In addition to forcing the operator to adopt a static position, intensive VDU work may encourage rapid and repetitive movement of the hands. Repetitive strain injuries (RSI) may be the result. MacKay (1985) observes that the number of terms used to describe musculoskeletal disorders and the imprecision of the terminology used in the literature have brought about a confusion as to what exactly are RSI. Bammer (1987) found that, in Australia, the term was used to encompass a large number of acute and chronic disorders of the upper body, whereas in most countries it is used to refer to problems in the lower arms, hands and wrists (such as tenosynovitis, peritendinitis and carpal tunnel syndrome).

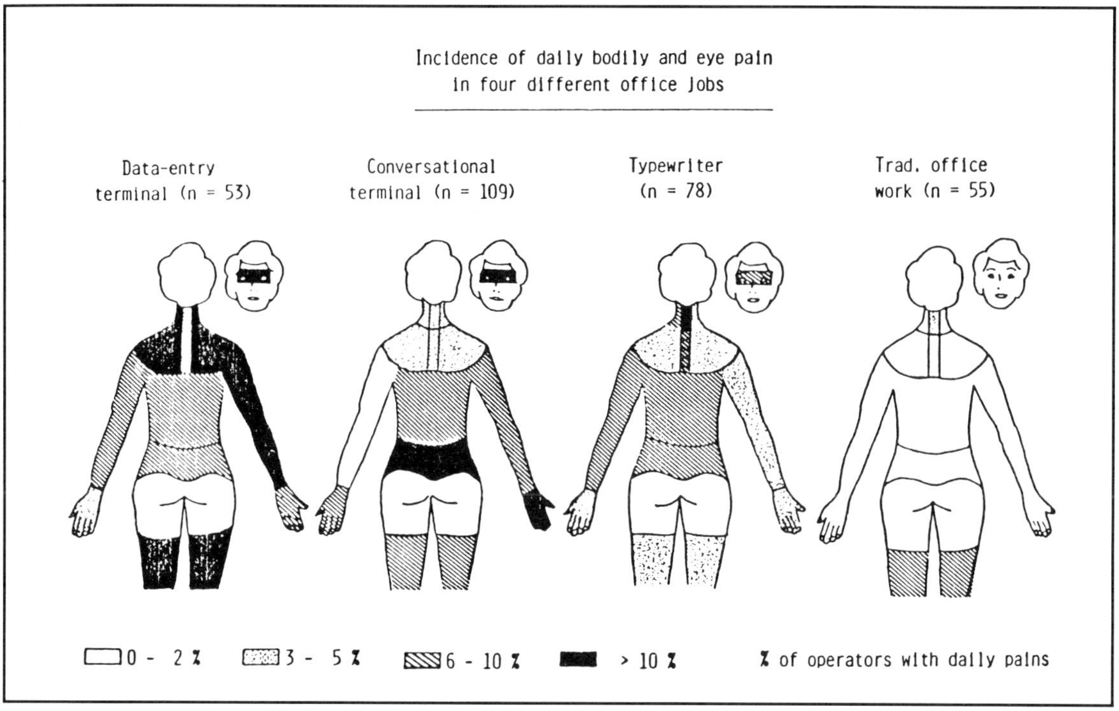

Figure 3.1 *Incidence of daily bodily and eye pain in four different office jobs (100% =* n *of each group).* Source: Hünting, (1984) © John Wiley and Sons, Ltd. Reprinted by permission of John Wiley and Sons, Ltd

Typically, symptoms may include numbness, pain, swelling, wasting of the muscles and loss of sensation in the hands and fingers. The *ILO Encyclopaedia of Occupational Health and Safety* quoted by the London Hazards Centre (1984, leaflet no.2, p.4) states of the treatment of tenosynovitis:

> In the early stages of tenosynovitis, treatment is simple; if the employees are instructed to report at the medical centre at the first sign of a sore wrist or forearm, infra-red ray treatment, tight bandaging and rest are usually effective. A change of work or job rotation is indicated to prevent recurrence. If . . . the employee [does] not report the injury until the pain is intolerable . . . medical or even surgical treatment is usually necessary.

Sadly, this advice would appear not to be taken as, according to Fosdal (1985), there have been a number of medical retirements due to tenosynovitis and, in addition, some employees are concerned that if they report a problem they may put their jobs in jeopardy.

The common connection between RSI is the likely cause of the symptoms, rather than the symptoms themselves. The cause is seen to be a combination of:

- static loading of the muscles, i.e. holding the body or parts of it still for long periods;
- repetitive movements of parts of the body;
- extreme force required by parts of the body.

Bammer (1987) argues that the move to VDU from manual systems, or even electric typewriters, may be to blame because the light keys encourage rapid keying, and because natural breaks caused by needing to change the paper, 'white out' or wait for the carriage return are eliminated. Thus VDU work can encourage repetitive movement and a static position.

The National Communications Union found that a major cause of RSI was the high productivity demands placed on the workers in question, resulting in rates of 10,000 to 12,000 keystrokes per hour (Hammond, 1986). Librarians would be forgiven for thinking that RSI would be less of a problem in libraries, where data input is rarely so intensive. As staff costs continue to increase, however, and budgets are reduced, the demand for greater productivity from existing staff will rise, and RSI may follow. Indeed, in one library in Australia, during a retrospective conversion project where targets were set, it was found that all six staff suffered from RSI. At a later stage in the same library, a new group of keyboarders was used, but no targets were set. Operators worked for 30 minutes at a session. No complaints were reported by this group, but the overall performance was three times as high as the first group (Ralli, 1988). This indicates that anxiety to achieve goals may not only cause unnecessary discomfort to the operators, but may also be counter-productive.

Hughes (1985) reported an increase of RSI among mature people, not trained as typists and starting to use keyboards when their joints were becoming less supple. A situation perhaps not untypical of many library staff.

Another cause of RSI was suggested by Professor Ferguson of Sydney University in 1974 (Hobday, 1985). Professor Ferguson asserted that it was the shape of the traditional Qwerty keyboard layout that caused the ill-health of Australian telex operators. (Keyboard layouts are examined in more detail in Chapter 9.)

The fact that RSI have common causes rather than common symptoms gives rise to two problems. The first is the difficulty in diagnosing RSI. Bammer (1987, p. 39) expresses the problem thus:

> RSI are a sea of unknowns. We do not know what the underlying pathophysiology is, how to treat them successfully, or how common

> they are. It is also not clear when and how discomfort develops into an injury. How much fatigue and discomfort is normal? How much pain should a worker be expected to live with? Is there a point when serious damage is done to muscles, joints and other tissues?

If it is not clear when damage occurs, should operators wait until the pain is too great to tolerate by which time damage may have occurred? In view of the comments by the ILO (above) this would seem unwise. On the other hand, it seems unnecessarily alarmist to rush to the doctor at the first sign of writer's cramp. Bammer (1987) found, in her visits to various countries, that VDU workers 'put up with' discomfort in the neck, shoulders, arms, hands, wrists and elbows by modifying some of their out-of-work activities. Usually workers cut down on home chores, sport and hobbies to give themselves more time to rest and recuperate. This is a situation echoed by the advice given in an article in the Australian Broadcasting Corporation's newsletter (*Electronic Library*, 1985, p. 15): 'muscles used for typing don't get much rest if you knit, play the piano, write or otherwise use your fingers in a competitive way'. For some, this may be better than having to give up work, but it is unfortunate that work should cause people to reduce leisure activities which they had hitherto enjoyed.

The second problem caused by the difficulty of defining RSI is that, according to Fosdal (1985, p. 115)

> . . . the Occupational Health Service refuses to recognise that anything other than conditions specifically on the DHSS prescribed list of industrial injuries has any connection whatsoever with keyboard work. They will, therefore, only recognize tensynovitis . . ., writer's cramp . . . . They refuse to accept that problems such as tennis elbow and frozen shoulder, which are extremely common among keyboard operators, are in any way related to keyboard work.

Musculoskeletal problems are typically short-lived – that is, they stop when the operator ceases working or shortly afterwards. However, 'it is unfortunately true that if people go on ignoring repetitive strain injuries . . . they can do permanent damage which is not reversible as we have seen from the cases of a number of our members' (Fosdal, 1985, p. 119).

How to minimize postural problems

To some degree, the various symptoms which may be experienced have already been grouped according to cause, but more specific culprits may be found. MacKay (1985) concludes that, overall, discomfort of the back and neck (occupational cervicobrachial disorders) is more common in full-time keyboard users than RSI (tenosynovitis, peritendinitis, carpal tunnel syndrome). He identifies the following causes:

- abduction of the hand (bending or twisting the wrist);

- Pronation of the forearm (turning the palm downwards), often close to the anatomical limit;
- raising the shoulders;
- lack of appropriate support for the back;
- prolonged work spells;
- lack of support for the hands and wrists;
- sharp edge of desk cutting into wrist or forearm

He also makes a number of recommendations to help to prevent postural problems:

- improvements in keyboard design (for the most part, out of the library's control, but for details see Chapter 9, 'The Keyboard', p. 91);
- altered work practices including job redesign (see Chapter 16);
- attention to postural requirements through workplace design (see Chapter 13);
- selection and training to improve or modify technique;
- occupational health care.

One library implemented a series of exercises for their staff during rest break periods and formed aerobic exercise groups (Ralli, 1988). The same library formulated a return to work programme for staff who had been suffering from RSI, which incorporated a variety of tasks and frequent rest breaks.

However, in the same way, that the causes of some physical symptoms cannot be distinguished, other causes compound the issue (Cakir *et al.*, 1980, p. 192):

> The problems of working posture, i.e. the positions and movements of the body whilst performing a given task, depend on the nature of the task in question, the physiological structure and the function of the body, and also on the psychological attitude and loading of the individual at work. It is essential, therefore, that when complaints of postural discomfort occur, the working situation should be investigated in its entirety in order to trace the most likely cause of the problem. In many cases, the problems will be found to lie in the poor design of the work place or working environment in which case they can usually be put right without too much difficulty. In others, complaints of discomfort might be the result of a more basic dissatisfaction at work, finding the solution to which might require a reappraisal of the organization of the task in question.

In addition, McKay (1985) argues that confusion can arise in that similar symptoms may be experienced with no bearing at all on causes in the workplace.

4 EYE AND VISION PROBLEMS

Introduction

Together with a rise in complaints about physical aches and pains occurring with the use of VDUs, has come an increase in complaints of visual discomfort. The following problems have been reported:

- Eye discomfort
  - soreness, grittiness, burning, itching, dryness, aching
  - tenderness around the eyes
  - redness of the eyes
  - irritation, flickering sensations
- Impaired vision
  - loss of visual sharpness
  - difficulty in focusing
  - double vision
  - blurred vision
  - seeing colour fringes
  - difficulty in looking at one point for a period of time

These may lead to other symptoms, such as headaches, and muscular aches in the neck, shoulders, back and arms (possibly from adopting a poor posture to obtain legibility of the screen), fatigue, stress or irritability. All these problems may disappear after a time, and have not been proved to cause any permanent ill-effects, but may be unpleasant while they last.

There is no proof of a causal connection between VDUs and visual discomfort, but this appears to be mostly due to the problems of scientific testing (that is, it is difficult to get a control group of non-VDU workers to perform tasks which are sufficiently comparable to VDU tasks). It would seem likely that factors such as a poor-quality screen (one which flickers and is difficult to read) or glare will give rise to some visual discomfort and reduce operator efficiency. This chapter discusses the features of VDUs which will detract from or enhance visual efficiency.

First, however, it is perhaps helpful to look briefly at how both the human eye and the VDU work.

How the eyes work

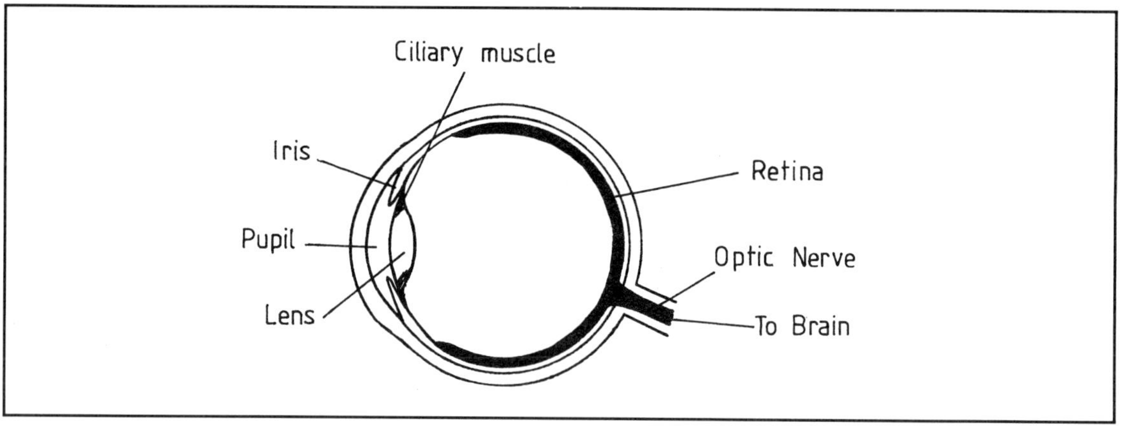

Figure 4.1 Diagram of the longitudinal section of the eye

Our eyes receive light waves through the pupil and these fall on the retina, where nerves receive and transmit the message to the brain. To assist in sighting and focusing on an object, we have three sets of muscles:

- One set of muscles moves the eyes from side to side and up and down.
- The iris muscles in each eye adjust the size of the pupil according to the varying light intensity. The pupil grows larger in dark conditions to admit more light and smaller if the light is bright to reduce the amount of light entering the eye.
- The ciliary muscles change the shape of the lens to focus the eye. When the eyes focus at close range, the ciliary muscles compress the lens. Close-up viewing for long periods therefore is very tiring to these muscles.

So tired eyes can be caused by:

- constant movement of the eyes (up and down or left and right), too little variety of eye movement and focusing;
- frequently changing brightness, causing the iris muscles to adjust the pupil size;
- close-up viewing for long periods or constant re-focusing.

Factors affection vision

Various factors affect our eyes:

Age

As people grow older the lens of the eye gradually loses its elasticity, and close-up vision becomes more difficult. Usually this is corrected by glasses, but if it is not, visual discomfort may occur whilst doing close work.

Lighting

The amount of light available also affects our sensitivity to certain colours. When it is dark, the eyes are more sensitive to blue than when it is light and, conversely, as it becomes lighter, the eyes become more sensitive to red. Green and yellow fall in the middle of the range. This in turn, therefore, affects our perception of colours on a display, green and yellow standing out more clearly under normal conditions than either blue or red.

The level of illumination will also affect our ability to read printed text and, as we get older, we need more light to read. However, it is also true that as age increases, so does sensitivity to glare.

The clarity of VDU and printed text needs to be of high quality under conditions of poor lighting or for older employees.

Different lighting levels are also needed for reading from a VDU screen than for reading written or printed text. Thus no single recommendation can be made, rather the three factors of operator's age, the source document and text on the VDU screen need to be considered together. The best way of doing this is to allow the operator to experiment with different lighting levels.

Glare

Glare can cause considerable visual discomfort especially as people get older and, in some cases, can even impair vision. It may be caused by light reflected from the VDU screen or the work surfaces or coming directly from a light source. Reflected glare, of the kind which occurs in libraries and offices, may cause discomfort but is unlikely to impair vision. Glare from direct sources, such as lights, should be avoided.

Glare is relative – that is, a light source may be bright compared to its surroundings and therefore cause glare. This can be reduced by bringing the overall level of lighting up to a similar level, but care must be taken to ensure that the overall level of lighting does not cause the screen contrast to be diminished.

The relative position of the glare source to the operator is also important; the closer it is to the centre of the visual field, the more performance is impaired. For more details about lighting and glare, see Chapter 10.

Pupil size

In order to adjust the amount of light received by the retina, the iris muscles adjust the pupil size. The length of time taken to do this varies and can sometimes be relatively slow. The pupil aperture is also affected, to some degree, by the focusing of the lens, contracting when the lens is focusing on near objects. Thus, good light is needed for close work because the pupil will not compensate in response to the light in the same way that it would for more distant objects.

Retina adaptation

The retina takes a long time (approx. 25–30 minutes) to adapt from light conditions to complete darkness. Adapting to light conditions is, in contrast, much quicker. However, when a bright light falls on the retina, overall sensitivity to light is reduced. Thus, if a VDU faces a bright light source, such as a window, the sensitivity of the retina to the light on the screen is reduced and reading is impaired.

Defective vision

Few people have 'perfect' vision; indeed, sight for most people deteriorates with age. It has been rumoured that constant use of a VDU can cause 'eye strain' and deterioration of vision. It is not possible to strain the eye muscles, although tiredness may result from any visually demanding task, such as proofreading, whether on- or offscreen. Even such intense work is more likely to reveal an existing defectiveness of vision than to induce it. It is therefore recommended that operators seek an eye test before starting any visually demanding task, including VDU work, and obtain corrective glasses if necessary.

People who wear bifocals may have particular problems because the part of the glasses intended for close reading is at the bottom of the lens. In order to view text on the screen, the operator may have to tip the head backwards in order to see through the lower part of the lens. If this is sustained for long periods or performed frequently, head, neck and shoulder pains may result. It may therefore be advisable to purchase a separate pair of glasses specifically for VDU work.

Contact lens wearers may also suffer, this time from dry eyes. People tend to blink less than normal when operating a VDU, with the result that the eyes may feel dry and uncomfortable. Eye-bathing or 'liquid tears' (which may be purchased at most chemists) may help (Doyle, 1988).

Reading

Our eyes are constantly on the move, taking in information from our surrounding environment. When we read, our eyes move along the text in short, jerky movements called saccades. The eyes jump from point to point, each point being approximately 4–8 letters away. At each point, the eyes come to rest and the word, or words, on which the eyes are focused are read. This is rightward reading saccades (presumably leftward on Arabic and similar scripts). Occasionally, the eyes will 'hop' back a stage to check what has just been read (correction saccades) and, at the end of a line, they swing left and downwards, to start at the beginning of the next line (line saccades).

To assist the reading process, it is important that characters are clear and easily distinguished. There must also be a sufficient gap between the lines to avoid confusion. If lines are too close together, it will take longer for the reader to distinguish the shapes (words) which are relevant and to pass on, hence reading speed will be reduced.

Character contrast is also important. If contrast is poor, the reading field is narrower and reading speed is reduced. Coloured letters, too, can cause a reduction in reading speed, although if they are well used they can improve the retrieval of specific pieces of information, such as by highlighting part of a record. Legibility is equally as important for the printed or handwritten source document as for the VDU screen.

How a VDU works

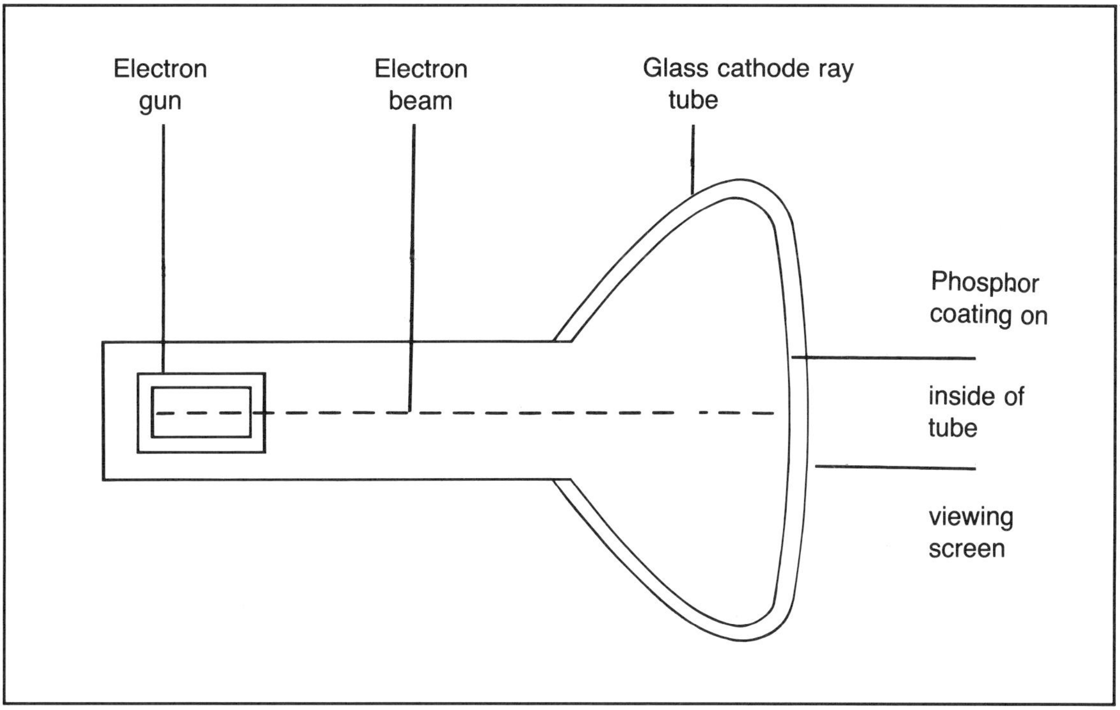

Figure 4.2 Simplified diagram of the longitudinal section of a VDU screen

Figure 4.2 shows a section through a VDU. The screen is the flattish part of the glass bulb on the right-hand side of the picture. The inside of the glass is coated with phosphors. The electron gun scans across and down the screen very quickly and, on receipt of a signal, usually from the keyboard, fires an electron beam at the screen. This beam activates the phosphor which then starts to glow, causing a visible image to appear on the screen. The phosphor is activated in dots and several dots together are used to form a character (see below).

The causes of visual discomfort

The problems listed above could arise from a number of factors relating to the VDU, the surrounding environment, and the type of work performed. The surrounding environment and its effects are considered in some detail in Part IV. The following is a discussion of the features relating to the VDU and the type of work performed as contributory factors to visual discomfort.

The VDU as a cause of visual discomfort

Several aspects of the VDU may be to blame:

Flicker and ghosting

Although the VDU image may appear to be still, it is in fact constantly changing. From the moment the electron gun is fired and the phosphor begins to glow, the glow starts to fade, until it is once more charged by a beam from the gun. The frequency with which the phosphor is restimulated is known as the refresh rate. If the refresh rate is too slow, the screen will appear to flicker.

The majority of VDU makes have a refresh rate of 50 or 60 Hertz (Hz) or cycles per second. The higher the rate, the more stable will appear to be the image. However, in this range (50–60 Hz) it is possible for some flicker to be perceived. One possible area of concern raised by Stewart (1980) is that many VDUs are designed for sale in the USA and are designed for USA voltage rates. When they are converted for UK mains use, the refresh rate falls to 50 Hz.

The refresh rate, however, cannot be considered in isolation from the persistence of the phosphor – that is, how long the phosphor stays illuminated after being charged by the beam from the electron gun. A number of different phosphors are currently in use, which all have different decay times (that is the length of time taken for the phosphorous glow to reduce to 10 per cent of its starting level). The slower the rate of decay, the less frequently the screen needs to be refreshed to avoid flicker.

A highly persistent phosphor can, however, bring about the problem of *ghosting* – that is, where the old image persists on the screen for a little while after new text has replaced it or after scrolling (movement) of the text has occurred.

Different people vary in their awareness of flicker and an individual's perception will be affected by his position relative to the screen. Flicker is more easily observed if the VDU is at an angle to the operator and thus seen out of the corner of the eye, than if viewed through the focusing part of the eye. One simple way of assessing if a VDU screen flickers is to hold a straight piece of card or paper under the text. If the characters appear to 'dance', the screen flickers.

Stability of the characters

If the electron beam is well regulated – that is, it travels across and down the screen in an orderly manner – the characters will appear stable; if not, they will appear to move. This movement may be defined in three ways:

- drift – that is, where a character gradually moves;
- jitter which, as its name suggests, is a jerky change of position of the character;
- disturbed linearity, which is where lines in graphs appear to move.

The contrast between the character and the screen background

People vary in their preferences for character brilliance and contrast, some preferring bright characters with high contrasts and others opting for dim characters and low contrasts. Factors such as age, the sharpness of the character, the level of lighting in the work environment and reflections from the screen will affect this. The less clear a character is, the greater the contrast needs to be between it and its background to maintain legibility. The surrounding lighting and any reflections from the screen will similarly affect the amount of contrast needed. It is important that users should have the facilities to change the brightness and contrast on their VDUs.

The above characteristics point to the need to take care when selecting hardware to ensure a clear and legible screen. Other factors affecting visual discomfort are to be found in the workplace itself.

Reflections on the screen

Reflections on the display screen are a cause of many complaints of visual discomfort. They reduce the contrast of the display, rendering it difficult to read, or even obscure the display altogether. Reflected light may also cause glare resulting in visual discomfort.

Ways of reducing visual discomfort

There are two factors which cause light to be reflected from the VDU screen. The first is the irregular surface of the phosphor and the second is the polished surface of the glass. Various methods of eliminating sources of reflected glare are discussed in Chapter 10, but if it is not possible to remove the sources of glare it may be helpful to consider the use of filter panels, screen coatings or VDU hoods, which have been designed to help reduce the problem.

Filter panels

Manufacturers have created a number of panels to fit over the front of the screen (see Figure 4.3) for a variety of reasons: to reduce glare; to change the apparent colour of the screen; to reduce the effects of static; or lead-impregnated screens to provide protection from alleged X-ray emissions. It is not possible, however, to put a filter over the display without reducing the brightness of the characters. Grandjean (1987) reports a study which concluded that, when a filter was used, the VDU needed to be operated on a higher brightness setting, which caused the phosphor to age more rapidly thus reducing the lifespan of the cathode ray tube (CRT).

Figure 4.3 A filter panel or anti-glare screen. Photo courtesy of Acco Europe Ltd

Filters vary in how they are made and the benefits they purport to offer. Some of these are described below. Many are offered by suppliers of computer equipment (see Appendix), the most common being the micromesh filter. Others, such as etched screens, are created during the manufacturing process.

- *Micromesh filters* are fine fabrics which are placed directly over the screen. Only light which falls perpendicularly on the mesh can penetrate and so reach the screen. Most light in the surrounding environment will fall at an angle to the mesh and thus fail to penetrate it, hence much reflected glare will be eliminated. However, the light emitted by the characters is also reduced by the filter, especially if the screen is viewed from an angle. The contrast between the characters and the background is reduced and, if the micromesh collects dust, the problem is further exacerbated. Cakir *et al.* (1980) conclude that 'Micromesh filters are an effective way of reducing reflections but not without certain penalties, some of which outweigh the advantages to be gained to such an extent that their use is often of questionable value', but this is contradicted by reports cited by

Grandjean (1987) which indicate a strong preference by operators for micromesh filters over other forms of anti-reflective device.

- *Neutral density filters* reduce the amount of light which passes through them. Light from various sources in the room is reduced on its way through to the screen and again on its return as reflected light. Light emitted by the characters only passes once through the filter and is therefore reduced less than that of the reflected light, so causing it to stand out more clearly. It is important, however, that the filter should be able to withstand the wear and tear of normal use and cleaning, as a scratched surface reduces the clarity of the characters.
- *Coloured filters* operate in the same way as neutral density filters, but they change the light which passes through them and hence the apparent screen colour. Studies have shown them not to be particularly effective (Grandjean, 1987).
- *Polarization filters* polarize the light passing through them and effectively reduce glare. Unfortunately, they are usually made of an extremely smooth substance called plexiglass which itself reflects light! It is, therefore, necessary for the filter itself to have a non-reflective coating. This kind of filter can be expensive.
- *Etched screens* have an irregular surface which reflects light in a variety of directions, thus reducing the amount which is transmitted back to the operator. This may be achieved by roughening the display screen glass either by chemical or mechanical means. However, some etched screens also disperse the light emitted from the phosphor and hence blur the characters. This is by no means true of all etched screens, however, and many effectively reduce glare whilst retaining clarity of characters. Etched glass placed in front of the VDU screen may be less effective because the greater the distance between the light source and the roughened surface, the greater the dispersion of the light. The distance between the characters and the etched glass will cause the clarity of the characters to be diminished. Interestingly, in a study reported by Grandjean (1987), a roughened screen display gave a better performance than either a micromesh filter or a coloured plastic filter.

Screen coatings

A variety of screen coatings are available which reduce reflected glare in a number of ways:

- *Spray-on coatings* deposit a rough film over the display screen, causing reflected light to be dispersed, but also reduce the definition of the characters.
- *Vapour-deposited coatings* work on a similar principle to spray on coatings, but create a finer surface with less detriment to character legibility. They are, however, more expensive.
- *Quarter-wave coatings* are so called because the deposited film has a thickness of a quarter of the wavelength of light. The result is that light passes through the filter but is inhibited on its return. However,

the sharpness of the characters is not diminished. The disadvantage of this coating, which is the same type as that used on good-quality cameras, is that they are easily soiled and particularly susceptible to fingerprints.

Tube shields/ hoods

Various shapes of CRT shields and hoods have been designed to reduce the glare, particularly from overhead lighting. These can be very effective, but can also have the drawback of causing the operator to adopt a rigid position.

In conclusion to this discussion on antireflective products, it is pertinent to quote Grandjean (1987, p. 86)

> All antireflective techniques have serious drawbacks. If efficiency is weighed against the drawbacks, the quarter-wave coatings and the etching/roughening procedures are preferable, whereas micromesh, polarization and coloured filters cannot be recommended. The adequate positioning of lights and the appropriate positioning of the screen with respect to windows remain the most efficient preventative measures [as well as the cheapest!]

Screen colour

Most VDU screens are monochrome, usually with green, white/grey, yellow or orange characters on a black background. Many operators have a strong personal preference for a particular colour. For the most part, scientists argue that there is no physiological reason why this should be so, although Osaka (1985) concluded that visual fatigue changed as the VDU colour changed. He found that blue and red, which are at opposite extremes of the visual spectrum, cause significant visual fatigue, when compared with green and yellow, which lie in the middle of the visual field. As monochrome screens use colours which are in the yellow–green spectrum, colour would not appear to be of significance in their use.

The increasing availability of colour screens, however, is encouraging software suppliers to experiment with a wide range of colours. This may in turn have some effect on the visual comfort of the operator.

Positive versus negative screen images

The majority of VDU screens have a dark background with coloured characters. For the purposes of this discussion, this will be referred to as a negative screen, while a positive screen will be defined as one with dark characters on a light background. (There is considerable discrepancy in the literature on this, with the majority of writers opting for the reverse of the above. The authors' justification for going against the stream is that print on paper preceded screens and represents to many people the positive image of dark on light.)

The negative screen requires office lighting that is a little lower than normal to ensure adequate contrast between the characters and the background. As the level of ambient lighting increases, the contrast decreases. Reduced lighting levels may, however, render the source document harder to read. Another problem which may arise from using a negative screen is that, in some tasks, the eyes must move constantly between the light (mainly white) source document and the dark screen. This may result in tiring the iris muscles as they continually change the pupil size to accommodate the varying amounts of light.

In contrast, the use of a positive screen brings no such problems of accommodation and, in addition, is not subject to the reflected light which is common with negative screens. However, it is not entirely free of problems. On a positive screen, the character width appears diminished because of the way in which it is created. The same dots are used as on a negative image, but the characters are formed by the absence, rather than the presence of light. The result is that the light dots surrounding the character tend to impinge on the character strokes, making them appear narrower than they really are. Grandjean (1987) quotes studies by Bauer, and Bauer and Cavonius which concluded that, for strokes to be perceived as equal in width, dark strokes must be 23 per cent wider than bright strokes. Thus the character set on a positive screen must be composed of sufficiently thick strokes to combat the slimming effects of the light background.

Positive screen images also are more susceptible to flicker than negative screens. This may be due to the white phosphor which is used which is not as persistent as some of those used on negative screens, and to the fact that there is a larger area of activated phosphor. Indeed, it has been reported (Grandjean, 1987) that a refresh rate of 92 Hz (or cycles per second) or 100 Hz is required to ensure a flicker-free positive screen.

Some research has shown that operators perform better with positive screens, feeling more comfortable and operating more efficiently (see below).

The legibility of VDU characters will affect the efficiency of operation and the visual comfort of the operator. Factors which affect legibility are described below.

Screen size Generally speaking, the larger the screen, the easier it is to read (although other factors also affect legibility). As libraries move towards the idea of full text retrieval, screen size becomes more important. High-quality large screens (that is, A4 or even double A4) are now becoming available and are highly recommended for reading large amounts of text on screen.

Character set In 'How a VDU works' (p. 37), characters were described as being composed of dots. This is something of a generalization as, in fact, there are four methods of creating characters on a VDU screen:

- facsimile generation in which a 'Master' letter is copied on to the screen;
- stroke generation in which the character is created by a series of strokes (as in writing a character);

- Lissajous generation in which the shape of the individual characters is built up by a series of segments of curves (lissajous curves);
- Dot matrix generation in which the shape of the character is made up of individual dots highlighted within a rectangular matrix.

The most common method is dot matrix generation.

Matrix size

The dots are divided into rectangular matrices, typically 5×7 or 7×9.

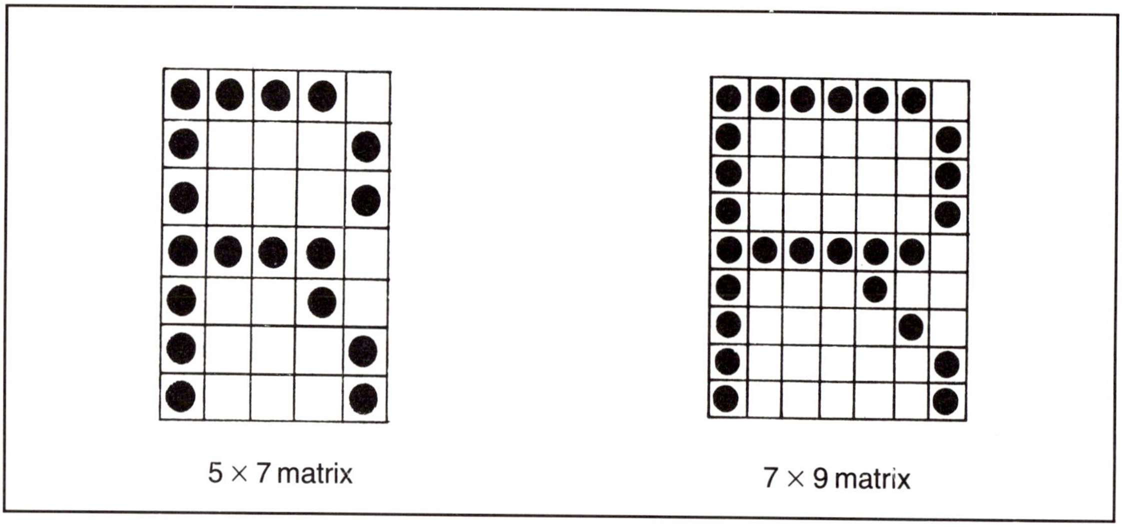

Figure 4.4 Two dot matrices used to create the letter 'R'

In Figure 4.4, two matrices, 5×7 and 7×9, are used to create the letter 'R'. The pattern on the right is clearer than the one on the left, because there is relatively less spacing between the dots. Because the right-hand image is clearer, it can be recognized more quickly. Similarly, on a VDU screen, the clearer the image, the faster the likely reading rate. Thus, the larger the matrix (within limits), the greater likelihood that the characters will be legible, and the more easily will the text be read by the operator. However, very large letters can reduce reading speed.

Character font

The character font which is used will also affect legibility. To some extent the font will be affected by the size and shape of the matrix. The more dots in the matrix, the clearer the image will tend to be and the more obvious will be the ascenders and descenders. Some fonts do not have proper descenders, the tail of the q or the y lying above the line. This makes the text harder to read. While it is important to have ascenders and descenders which are in proportion, serifs only hinder the reading process as they tend to clutter the screen image. Slanting characters are also harder to read than an upright Roman script, and capital letters exclusively are less legible than a mixed case font.

Whatever font is used, it is possible for confusion to occur between the characters. An 'I' for example may be misread as '1' (one) or vice versa. Knave (1983, p. 32) lists the following characters as possible candidates for confusion. The first column 'mutual' contains those characters which may be mistaken for each other. In the second column, 'one-way', it is more common for the characters on the left to be mistaken for those on the right than vice versa. The entries marked with an asterisk (*) are those characters which together cause more than 50 per cent of the total confusion:

Mutual		One-way		
			mistaken for	
O	Q	C	"	G
T	Y	D	"	B
S	5	H	"	M, N
I	L	J, T	"	I
X	K	K	"	R
*I	1	*2	"	Z
		*B	"	R, S or 8

When selecting VDUs, therefore, it is important to consider character font.

Character size

The legibility of the information on the screen will be affected by the size of the characters used. Appropriate character size is dependent on viewing distance, among other factors, so that, for example, the text of a viewdata system should be large enough for visitors to the library to read easily at a glance when standing at a distance of, say, three metres. Text intended for close-up reading, however, can use a similar character size, with the added advantage of fitting more text on a single screen.

The height-to-width ratio also affects legibility. Characters which are tall and thin or short and fat are harder to read than those which are of squarer proportions. Other factors affecting character legibility are: character contrast; the dot matrix used; lighting; and, of course, the operator's eyesight.

For a viewing distance of between 50 and 70 cm the recommended proportions for characters are as follows (Grandjean, 1987):

Height of capital letters	3–4.3 mm
Width of capital letters	75% of the height
Height of lower case letters	70% of the height of capital letters
Width of lower case letters	60% of the height of capital letters
Stroke width	20% of the height
Intercharacter distance	25% of the height
Interline space	4–6 mm or 100–150% of the height of the capital letters

For displays, such as viewdata, which are more commonly viewed from a distance, character size needs to be larger but the relative proportions should remain the same. Smaller characters than those recommended are likely to result in reduced legibility.

Cursor

The cursor is essential for interactive work to indicate the position at which data are to be entered. Vartabedian, as quoted by Knave (1983), lists five qualities necessary for a cursor. A cursor should:

- be easily found in random location in an alphanumeric display;
- be easily tracked as it moves through the display;
- not interfere with the reading of symbols;
- not impair search time for non-cursor related information;
- not use a graphic technique or shape used elsewhere in the display.

Cursors come in a variety of forms, some of which may be altered with the terminal. The most common are a long underline (_) and a box (□) in inverse video. After studying a variety of cursors, Vartabedian found that a box cursor flashing at 3 Hz was optimal.

Interline spacing

The gap between the lines of text on a screen affects reading performance. It is recommended that a minimum interline distance of 1/30 of the line length should be used; thus as the lines increase, so should the interline spacing. Grandjean (1987) concludes that it would be advantageous to use vertically oriented screens enabling shorter lines and hence smaller interline distances to be used. This would accommodate more text per screen.

Character sharpness/ resolution

It is not just the size and shape of a character which is important, it is also the clarity which is necessary for legibility. A sharply defined character will be easier to read and will thus give a higher reading performance and reduce visual discomfort. Some VDU characters have a blurred edge which minimizes the clarity of the character. If the blurred edges are wide relative to the width of the stroke which forms the character, then the blurring will be clearly visible.

Screen display

This is discussed in more detail in Chapter 15 on the software interface, but it is important to know that, even if the VDU screen offers all the correct features for legibility, poor screen display brings its own problems. The Civil and Public Services Association Research Department (1983) reports the use of The Graphix Optical Character Recognition System. The system was installed in the DHSS in 1976 to perform data entry tasks and ran so successfully that bureau work was undertaken, in this particular instance to do cataloguing work for the British Library. The process involved the system displaying all the text which has been read in and by highlighting text that has not been recognized. Unfortunately, British Library data contained a great deal of text that was unrecognizable to the system, including the Cyrillic alphabet. Hence, large proportions of the screen were highlighted at any one time resulting in operators being forced to look at a bright screen for long periods and to correct the brightly lit, unrecognized, text. As a consequence, fatigue set in quickly (after about 20 minutes) and frequent rest breaks were required.

The type of work performed

Different aspects of VDU work impose their own visual demands. Some of these are considered below:

The task performed

The more visually demanding a task is, the more tiring it will be. Some tasks are more visually demanding than others. Online cataloguing, for example, may require the operator to search a full MARC record to locate relevant information. This is visually more intensive than the use of a brief catalogue record. During such a task, it is desirable for the operator to be able to rest his eyes from time to time – for example, by looking at a distant view or a picture.

The source document

Some tasks, such as word processing, require much concentration on a source document. If its characters are poorly formed and difficult to distinguish, or if the page layout is poor, reading speed will be diminished as, too, may accuracy. Other factors which affect the readability of the source document are the reflectivity of the paper surface, the colour of the paper and the colour and density of the ink.

The angle of the light on the source document also affects its readability. If the paper is at right angles to the light source, the white background will appear bright in contrast to the dark letters. As the light moves relative to the source document, the contrast becomes less marked and the text, therefore, less legible.

Reading from a VDU

Much work has been done comparing the performance of people reading text from a VDU screen with the same people or others reading from hard copy. Gould and Grischkowsky (1984) found that reading from a VDU screen is 20–30 per cent slower than reading the equivalent text from hard copy. They also found that although changes to vision, performance and operators' feelings occurred throughout the day, these were due to the task performed and were not affected by the use of a VDU. In contrast, Cushman (1986), studying the effects of reading text for 80 minutes using microfiche, VDUs, both negative and positive images and printed paper copy, found that visual fatigue was significantly greater when subjects read from negative (light on dark) microfiche or a positive (dark on light) VDU than the other media. This indicates that there may be some problems associated with reading from a VDU screen for any significant period, but it is by no means conclusive as the quality of the VDU screens used may have been at fault. Indeed Gould, Alfaro *et al.* (1986) reported that reading speeds equivalent to those on paper can be achieved by using high resolution, positive VDU screens with character fonts which resemble those on paper (i.e. that is, which are not formed by dot matrices).

Scrolling screen

Online searching may necessitate reading information from a scrolling screen. Far from being a cause of visual discomfort, it has been found to present few problems to the operator. Sekey and Tietz (1982), as cited by Hamdy (1987), found little difference in observing eye movements between reading a static display and scrolled text, especially when the scrolled display was rapidly paced. Hamdy also cites Kolers, Duchnicky and Ferguson (1981) who found that scrolling could actually assist more effective reading, with the operator making fewer and briefer glances at the text. They found an increase in reading speed with no apparent loss of comprehension. As the rate of scrolling gradually increased, performance correspondingly increased to the point where operators performed best at a speed which was 20 per cent faster than the pace at which they felt most comfortable.

Summary

Although there is no proof of a causal correlation between VDU work and visual discomfort, there is a strong indication. In a colloquium organized by the French Society of Ophthalmologists (Boyle, 1986), the following factors were cited as the principal causes of dizziness, prickly eyes and headaches affecting a claimed 70 per cent of users:

- frequent eye accommodations – both movements between screen, keyboard and paper (with an estimated 30,000 such movements being needed during a working day) and the constant adjustment between positive and negative contrasts;
- screen image flicker (refresh rate);
- dazzle effects caused by poor lighting;
- screen reflections;
- poor character readability.

Not surprisingly, the group recommended that ophthalmic examinations should be made on the employees before starting to work on VDUs and that these should be repeated regularly, especially when the user is in his mid-forties.

The American Optometric Association (Nelson, 1984) recommends further steps:

- Use proper furniture and position reading material at a convenient distance; adjust the VDU screen so that it is slightly below the viewer's straight ahead seeing position and approximately 14 to 20 inches from the eyes.
- Place reference material as close to the screen as possible to avoid frequent large eye and head movements, and at the same distance

from the eyes as the screen to avoid having to change focus when looking from one to the other.

- Adjust lighting in the room for VDU use to less than half the customary office lighting level. (It should be noted that this advice pertains only to negative screens.)
- Reduce glare on VDU screens by placing terminals so that sources of bright light are not behind the operator; small hoods can be attached to extend above screens to shield them from overhead light.
- Give workers frequent rest breaks: the National Institute of Occupational Safety and Health recommends a 15-minute alternate task break after two hours of continuous VDU use for workers under moderate visual demands, and a 15-minute break every hour for those under high visual demands.

In addition, it should be noted that high refresh rate, positive screens with good character sets are to be preferred wherever possible.

For further information on how to adjust the work environment, see Part IV; and for information on job design, see Chapter 16.

5 STRESS

Introduction

Stress is commonly understood to be a work-related health hazard. Some employers now routinely examine their staff for early warnings of impending heart disease or strokes which are known to be partly caused by stress. For the most part, though, little is done in the workplace to reduce the cause of stress on people's lives. Indeed, in the US, the National Association of Working Women, which has undertaken considerable research into occupational health, found that the following beliefs about stress existed (Apex, 1985):

- People with the greatest responsibility, who make a lot of important decisions, have most stress.
- People bring stress with them from home into work and, if they are under stress, it is because of family or personal problems.
- Certain people are more susceptible to stress; this is not due to the job, but to the individual.

A common misconception seems to be that stress only affects important, highly-paid jobs and that it is, in a sense, the price the executive pays for the kudos of his career. Sadly, stress does not confine itself to a set of well-paid individuals, but often also affects those who are at the bottom of the ladder.

Nevertheless, we cannot conclude that automation necessarily produces stress; indeed, in many cases, the reverse may be true. Our concern here is to identify the areas where automation may create stress for some individuals.

What is stress?

Stress is our bodies' biochemical response to an outside stimulus. Different stimuli will cause us to react differently; we may blush, jump or hide the reaction under a confident mask, but the biochemical response of our bodies is the same – an increase in the heartbeat and breathing rates, increased secretion of stomach acid and the release of certain hormones. An increased flow of adrenalin and a rise in cholesterol and blood sugar cause a surge of energy to the muscles, whilst the stomach becomes inactive. Emotionally, we may experience frustration, anger, excitement or anxiety.

Some stress is good. Without thrills or excitement, life would be dull. Indeed, after a short period of stress, our bodies return to equilibrium. It is stress which is prolonged and continuous that poses a threat to health, as our bodies remain poised for 'fight or flight' and the resultant imbalance can wear our bodies down.

Stress and health

As a result of prolonged stress, a number of health problems may occur. Coleman (1978) cites no fewer than 194 symptoms and diseases which may be stress-induced. The following are the more common problems:

- Short-term stress symptoms
  - Indigestion, loss of appetite, nausea, vomiting
  - Irritability, restlessness, behavioural changes, boredom
  - Nervousness, anxiety
  - Chest pains
  - Insomnia
  - Headaches, migraines
  - Backache
  - Toothache (from grinding one's teeth!)
  - Susceptibility to colds and 'flu
  - Menstrual irregularities
  - Sexual difficulties
  - Fatigue, memory blanks
  - Depression, low self-esteem
  - Accidents
- Long-term stress symptoms
  - Ulcers, bowel disorders, spastic colon
  - Vaginal and bladder infections
  - High blood pressure
  - Coronary heart disease
  - Hypertension
  - Chronic mental health problems, nervous breakdown
  - Decreased resistance to chronic disease

The various symptoms are not inseparable, and may even create more stress for the person, thereby exacerbating the situation. They may also cause the sufferer to increase his consumption of tobacco, alcohol and drugs in order to relieve the symptoms. This, in turn, can compound the health problems.

The causes of stress

Work-related stress caused by the introduction of new technology can be divided into four main groups: environmental factors; job design faults; employee–employer relations (Apex, 1985); and the nature of VDU work. Two of the factors – the working environment and job design faults – are examined in Parts IV and VII respectively. Detailed consideration is given here only to employee–employer relationships and the nature of VDU work.

Environmental factors

A poor working environment may cause or add to job stress. Recommendations for improving the environment and workstation layout are made in Part IV.

Job design faults

Poor job design, resulting in an incorrect pace of work or underutilization of skills, can cause stress.

Employee – employer relationships

The introduction of an automated library system changes the traditional function in the library and the roles which each employee is expected to play in the system.

> Fundamental organizational change evokes feelings of apprehension at all levels . . . Uncertainty surrounding the need for one's own job in the new system or its possible restructuring with consequent changes in pay, promotion and training, etc., inevitably lead to anxiety. (MacKay and Cox, 1984).

Social isolation

Library work does not always require intensive use of VDUs, and yet it can bring with it a sense of isolation. For example, although the Browne issue system is cumbersome and time-consuming, some library assistants we interviewed remarked that it had given them the opportunity to talk to the user whilst searching for the book card. Automated systems, they felt, encouraged the assistant to 'process' the reader more rapidly and reduced personal interaction.

Failure to solve grievances

This may occur because management does not respond or there is a failure to understand the underlying causes and thus treat only the symptoms. Other problems may arise because the solutions may be seen to be costly to implement: finances for suitable furniture, for example, may be difficult to justify in a library that has only just won a hard-fought battle for an automated system.

Fear of adverse health effects

Unless the fear of potential health risks is handled sensitively and knowledgeably by management, it can increase out of proportion.

Threat of job losses

Automation is sometimes introduced to cut the staff budget, which may imply that existing staff are to be axed. Even if this is not the case in an individual library, the plan to automate may create anxiety.

The nature of VDU work

Professor Frank Zingrone of York University, Toronto, has propounded the theory that the nature of the VDU and of the tasks performed on it cause conflict between the right and left hemispheres of the brain (Else, 1985). The left side of the brain controls language and logic – skills which are frequently called into play during VDU work. The right side of the brain controls visual spatial capacity, imagination and intuition and thus the viewing functions at the VDU. This conflict, Zingrone suggests, causes stress, but this has been questioned by other researchers. He suggests no easy solutions, but recommends only short spells (20 minutes) on the VDU at any time.

The need to reduce stress

Unlike postural effects, stress effects do not go away when the person has stopped working; rather, they may have an insidious influence on the rest of the person's life – for example, increasing the time needed to wind down at the end of the working day. As this will obviously have a direct effect on the individual's private life, work-related stress can have a bearing on an employee's quality of life.

Similarly, not all stress is work-related. Holmes and Rahe (1967) formulated a table of stress points which are related to various lifestyle changes (see Table 5.1). The higher the number of points, the greater the level of stress, and the points are cumulative, so that if there is more than one stress-giving situation in a person's life, the points related to each should be added together to give the total stress level. More than 100 points at any time is likely adversely to affect the health of a person. Thus it is important to see that work-related stress will add to the pressures experienced elsewhere. It should be noted that this table relates only to stress caused by changes in people's lives; other stressful factors are excluded.

The adverse effects of stress will impinge upon the running of the library. Unnecessary tensions may be created, staff – user relationships may deteriorate, inaccuracies may develop in work, and so on. More serious effects could include employee absenteeism, which in turn would increase the load on fellow workers. Consequently, library managers should be aware of the need to reduce work-induced stress.

How to reduce work-related stress

Not all stress is a problem. As Cakir *et al.* (1980) affirm, the ideal is not the absence of stress, but that the amount of stress should be optimized. Not all people will have the same reaction to stress, since several factors will be involved.

- The individual's personality. A person's ability to cope is affected by his emotional stability and his degree of introversion or extroversion.
- His motivation. Some people are driven by the prospect of increased responsibility or a greater financial reward or other factors (see 'Motivation', p.7).
- The relevance of the stress to the work being undertaken. As a reference librarian, the challenge of a difficult enquiry may be

Table 5.1 Social readjustment rating scale (adapted from Holmes and Rahe, 1967)

Death of spouse	100
Divorce	73
Marital separation	65
Jail term	63
Death of close family member	63
Personal injury or illness	53
Marriage	50
Fired at work	47
Marital reconciliation	45
Retirement	45
Change in health of family member	44
Pregnancy	40
Sex difficulties	39
Gain of new family member	39
Business readjustment	39
Change in financial state	38
Death of a close friend	37
Change to different line of work	36
Change in number of arguments with spouse	35
Large mortgage or loan	31
Foreclosure of mortgage or loan	30
Change in responsibilities at work	29
Son or daughter leaving home	29
Trouble with in-laws	29
Outstanding personal achievement	28
Spouse begins or stops work	26
Begin or end school or college	26
Change in living conditions	25
Revision of personal habits	24
Trouble with boss	23
Change in working hours or conditions	20
Change in residence	20
Change in school or college	20
Change in recreation	19
Change in church activities	19
Change in social activities	18
Moderate mortgage or loan	17
Change in sleeping habits	16
Change in number of family get-togethers	15
Change in eating habits	15
Holiday	13
Christmas	12
Minor violation of the law	11

considered an enjoyable part of the job and thus enhance job satisfaction. The need to learn how to operate an automated system or to cope with computer failure may be seen as outside the person's remit and thus become a negative source of stress.
- The employee's attitude to the previous system. Some library assistants interviewed remarked that the laborious manual issue systems coupled with long queues of readers created more stress for them than learning and operating the automated issue system.
- The employee's attitude to the new system. If the previous system caused problems for him, he may welcome the automated system. Alternatively, he may prefer to stay with a system he knows and resent the change.
- Other stress effects in his life.

Thus staff will respond differently to the process of automation. To one, it may be a stimulus, to others a burden. Consequently, there is no single answer to the problem. However, there are a few pointers which may help to avoid, or reduce, stressful situations:

- Optimize the environment and workstation layout (see Parts IV and V).
- Ensure good job design (see Chapter 16).
- Communicate with all staff, telling them what is happening and explaining their role within the new system; also listen sympathetically to their comments, complaints, and fears.
- Where possible, sort out their grievances. If a solution is not possible, for financial or other reasons, tell them why.

6 REPRODUCTIVE HAZARDS

Introduction

The press is full of stories of 'clusters' of birth abnormalities amongst female VDU workers and some cases of sterility amongst men. The finger of accusation has been pointed at X-rays and at Extremely Low Frequency Radiation (ELFs) emitted by VDUs. In defence of the VDU have come equally numerous articles reassuring the user that all is well. The following is a summary of the discussions to date.

Reports of adverse pregnancy outcomes

A number of 'clusters' of miscarriages and birth abnormalities have been cited since the late 1970s (see Table 6.1) when VDUs came into use by different groups of people. VDUs were no longer the domain of programmers and systems personnel, but were now beginning to be used intensively for data input, by a group of users who were largely female.

With such a list of reproductive problems, it is hardly surprising that VDU work became a controversial issue. However, the reported clusters cannot be taken as proof that the VDU is to blame; the sample sizes are too small and the timescales are, for the most part, too short for the clusters to be of statistical significance. In addition, there may be other variables which are not taken into account, such as the environmental conditions, the stress induced by the work, the social class of the women and their general level of health. It has also been argued that many clusters are to be expected on the basis of chance alone. Bergqvist (1986) gives more detailed comments on some of the clusters.

Abnormal pregnancy outcome is an emotive issue, however, so it is unwise to leave the matter as 'unproven'. In order to confirm or deny the allegations, a number of studies have been, and are being, undertaken.

One of the most widely reported is the 'Runcorn' study, carried out by the Civil Service Medical Advisory Service (CSMAS) at the Department of Employment offices in Runcorn, Cheshire. Of the 803 female data processors studied, 36 per cent of pregnant women who had been

Table 6.1 Reported clusters of pregnancy problems (adapted from Health at Work, 1984)

Date	Location	No. Pregnancy Problems/ No. Pregnancies	%	Details	Estd. Level of VDU Use
1977–78	A bank in Grimsby	5/5	100	3 miscarriages, 1 stillbirth, 1 deformed baby	low
1978	Bank of Adelaide, Australia	–	–	4 miscarriages	low
1978–84	Public Library	8/10	80	8 miscarriages	low
1979–80	Toronto Star newspaper	4/5	80	4 congenital birth defects	high
1979–80	Sears Roebuck, Dallas, Texas	8/12	67	7 miscarriages, neonatal death	?
1979–80	Defence Logistics Agency Office, Atlanta, Georgia	10/15	67	7 miscarriages, 3 severe birth defects	medium
1979–81	Air Canada check-in counter	7/13	54	7 miscarriages	medium
1979–82	Solicitor General's Office, Ottawa	7/8	88	4 miscarriages, 1 premature, 2 born with respiratory diseases	medium
1980–81	Ontario Ministry of the Attorney General	10/19	53	10 miscarriages	medium
1982	British Columbia Employees Union	5/6	83	No details given	high
1983–84	Greater Manchester Fire Service computerized control	8/10	80	2 miscarriages, 3 hospitalizations with high blood pressure, 4 threatened miscarriages 4 emergency sections and problems at birth (some overlap)	high
1983-84	British Telecom Data Processing, Bristol	4/5	80	4 miscarriages	high

Note: Normal rates for miscarriage are in the 8–20 per cent range and, for birth defects, 1–4 per cent (Bergqvist, 1986).

exposed to VDUs for three months before and three months after conception experienced miscarriages, stillbirth or malformation. This compared with 16 per cent in a control group of women not working with VDUs.

Despite the apparently important findings, according to the CSMAS, the study was inconclusive for the following reasons (London Hazards Centre, 1984).

- The design of the study was unsatisfactory.
- The statistical analysis provided was inadequate.
- The control group reported an unusually low rate of pregnancy.

In addition, an independent analysis of the CSMAS report by Professor Geoffrey Chamberlain of the Department of Obstetrics and Gynaecology at St George's Hospital Medical School stated that 'one could not draw conclusions' (London Hazards Centre, 1984).

Brix and Butler (1986) analysed a number of other studies but found no strong association between VDU work and adverse pregnancy outcomes, either because the study found no such association, or because faults or biases were present. In their own study, Butler and Brix (1986, p.77) drew the following conclusion:

> It is possible that only extensive exposure to VDTs is associated with an excess risk and that less regular [frequent? ed.] use [i.e. less than 20 hours per week] of VDTs is associated with no excess risk.

This may be a comfort to library personnel, who were identified as low users of VDUs in the paper by Ericson and Kallen reported by Brix and Butler (1986). However, as more and more aspects of library work are automated, some areas, such as cataloguing, acquisitions and serials control, could become intensive in large or centralized libraries. Butler and Brix (1986) concluded that there was a possibility of excess risk of adverse pregnancy outcome in highly exposed workers, but that this was not proven. Even if there were such a risk 'it is less than the excess risk of other environmental exposures, such as alcohol and tobacco'. [p.77] More studies are underway and their results are awaited with interest.

No firm evidence is therefore available to acquit or condemn the VDU, but in the meanwhile let us consider the factors that may explain any link between VDUs and adverse pregnancy outcomes, should such a link be found to exist.

The alleged causes of reproductive hazards

The following factors have been put forward to explain the occurrence of clusters of birth abnormalities.

- Radiation and other emissions
  - X-ray radiation
  - Ultraviolet radiation (combined with an excess of vitamin A)

- Microwave and infrared radiation
- Low frequency electromagnetic fields
- Electrostatic fields in combination with light air ions
- Ultrasound
- Polychlorinated biphenyls (PCBs)

- Other factors
  - Stress and worry
  - Sedentary work, postural problems

Radiation and other emissions

Radiation is a beam of energy. Electromagnetic radiation is electrical energy in two parts, electric and magnetic, travelling in the form of waves. These waves differ in frequency (the number of times they go up and down every second). They also vary in wavelength (the distance from the peak of one wave to the peak of the next). Frequency and wavelength are related; the shorter the wavelength, the higher the frequency. Frequency is usually measured in Hz or cycles per second (see Figure 6.1)

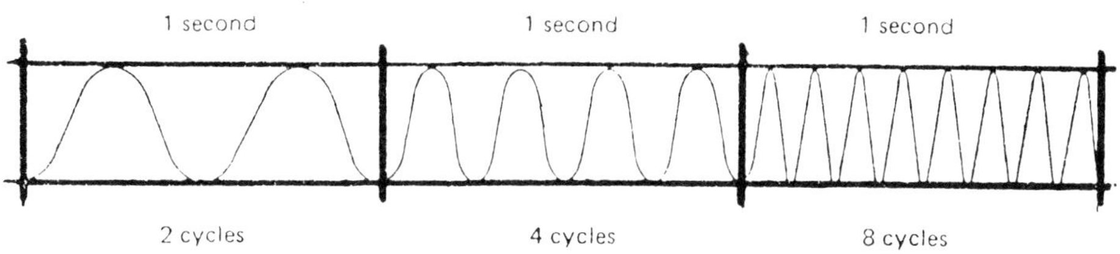

Figure 6.1 The pattern of radiation, indicating the inverse relationship between frequency and wavelength Source: Apex, 1985

Thus radiation, in its various forms, stretches along the spectrum. At one end are long waves with low frequencies and at the other, short waves with high frequencies. The higher the frequency, the more powerful the radiation (see Figure 6.2).

In the low-frequency end of the spectrum, the two parts of the wave – the electric field and the magnetic field – begin to separate. Different sections of the radiation spectrum have been highlighted from time to time as being the cause of birth abnormalities. The various types of radiation and their alleged effects are discussed below. The first to be accused was X-ray radiation.

X-rays

- Where do the X-rays come from? No gamma or hard X-rays are produced by VDUs. Soft X-rays are generated inside the cathode ray tube as the electrons from the gun (see page 37) are slowed down by the screen phosphor. The glass of many tubes has a high lead

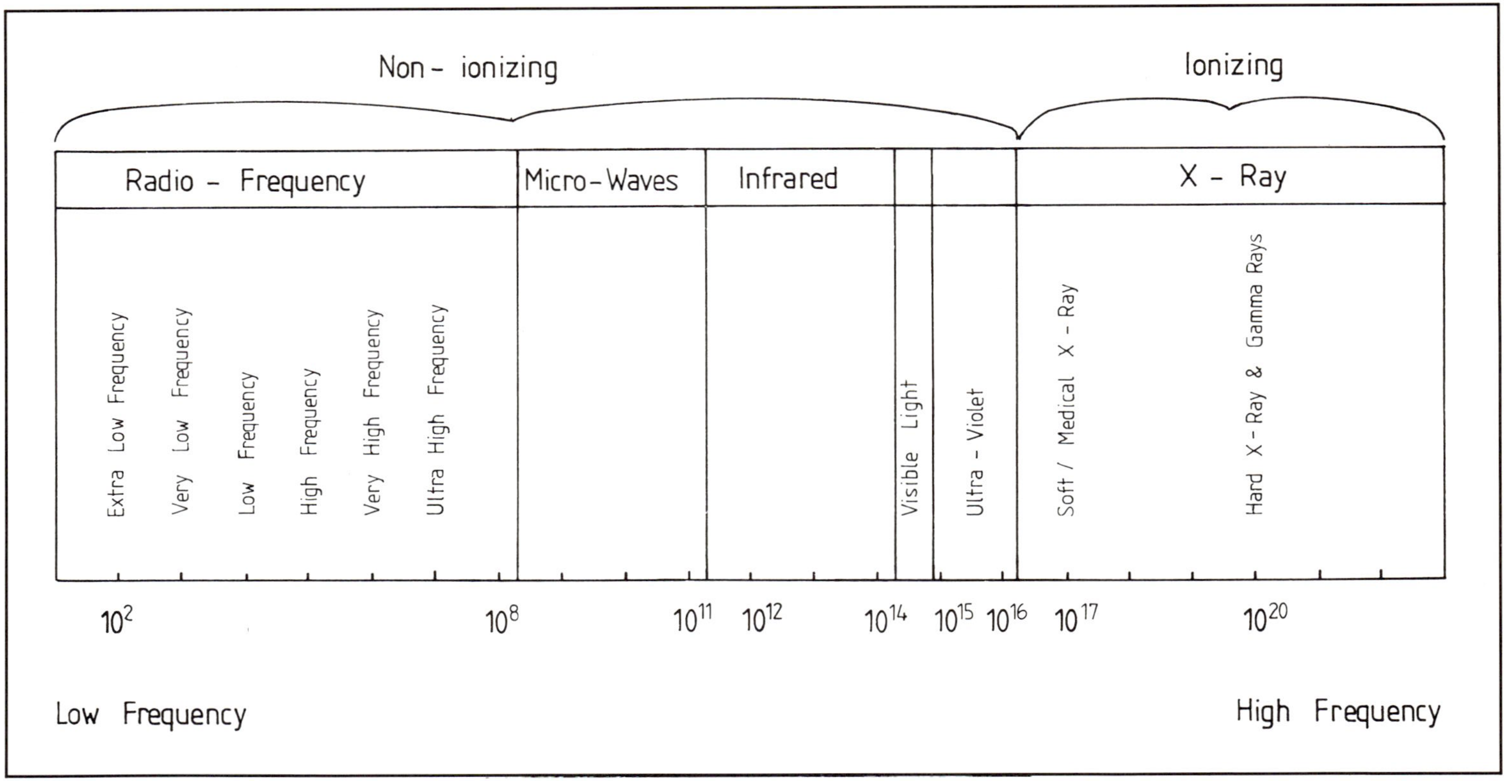

Figure 6.2 Schematic diagram of radiation

content, which blocks the passage of X-rays, and in addition the number of X-rays produced may also be reduced by correct circuit design.

- What harm could they cause? The effects of X-rays are cumulative and irreversible, so it is the total level of absorption rather than the level of emission which needs to be considered. If a person absorbs more than a certain amount, X-rays can cause cancer, genetic damage and other problems. They are also known to affect cell growth, and could therefore cause abnormal cell reproduction and harm embryos.
- What are the safety standards? In order to allow for the cumulative effect of X-rays, the measurements include an allowance for quantity and time (milliroentgens per hour or mR/h). Distance from the source of an X-ray also makes a difference – for example 100 mR/h at one foot would be reduced to 25 mR/h at two feet (Apex, 1985). Based on an assumption of a five-day, 40-hour week, official regulations for occupational safety in Europe and the USA set a maximum exposure of 2.5 mR/h at a distance of 40 cm from the screen (Grandjean, 1987).
- Do VDUs emit X-rays? X-rays were originally a cause for concern because some computer equipment from the 1950s was found to emit this type of radiation. Measurements of large numbers of more modern CRT equipment in use, however, have indicated that such emissions no longer occur (Bergqvist, 1986), although recent cases of radiation leakages have been recorded under test conditions (Apex, 1985). Bergqvist argues that the test methods themselves often include creating a faulty condition in order to examine the CRT when faulty as well as when operating correctly. He argues that X-rays emitted under these conditions are not a fair assessment of the VDU in a normal operating environment.

Grandjean (1987, p.195) cites the conclusions of the US National Academy of Sciences following a number of studies:

> Taken collectively, these studies have examined a wide variety of models and hundreds of terminals. Measurements of emissions from older and newer VDT models have not differed significantly. Measurements have been made under normal operating conditions and under conditions designed to maximise potential emissions. . . . These studies have concluded that the levels of all types of electromagnetic radiation emitted are below existing occupational and environmental health and safety standard limits of exposure . . . The level of X-ray radiation emitted by VDTs is far less than the ambient background level of ionizing radiation from natural sources to which the general population is exposed.

The last part was borne out when one of the authors' VDUs was given a routine check. It was a concrete pillar in the office, *not* the VDU, which caused the meter to react! From the studies undertaken so far, it is likely that VDUs do not emit X-rays and, if X-rays are emitted, these fall below the safety standards.

- How would I know if my VDU was leaking X-rays? It is possible to arrange for equipment to be tested periodically for unwanted emissions; a health and safety officer can advise. But perhaps you need look no further than the screen itself. John Bushman, quoted in *Mind Your Own Business* (VDU users – read and relax, 1986, p.69) reassures us thus:

> One thing X-rays do is destroy static. As any VDU user knows, the front of a cathode tube has lots of static; the point here being that if there were any significant X-ray emission, the static would not be present.

Perhaps now, we will stop complaining at the amount of dust which is attracted to the VDU screen!

Ultraviolet radiation

Large amounts of ultraviolet radiation are emitted by the sun, most of which is absorbed into the atmosphere. What remains is beneficial to life on Earth, helping growth and producing vitamin D in the body. Excessive doses can, however, lead to sunburn and may cause cancer.

It was suggested by Mawson in 1985 that a combination of ultraviolet radiation from VDUs and an excess of vitamin A, could be the cause of a large number of adverse effects from VDUs, including adverse pregnancy outcome (Bergqvist, 1986).

The amount of ultraviolet radiation emitted by VDUs is, however, negligible compared with that from fluorescent lights, and sunshine both outdoors and through glass. Therefore, if ill-effects were to be caused by a combination of ultraviolet radiation and vitamin A, it is likely that a source of ultraviolet radiation other than VDUs is to blame.

Microwave and infrared radiation

These non-ionizing rays can cause heat to be generated in the substances absorbed by them, hence the effectiveness of microwave ovens and infrared heating appliances such as electric blankets.

Concern exists about the heating effects of these rays on the body. This has been caused by reports of sterility and sexual problems among men who have been exposed to microwaves. The testicles are particularly sensitive to heat, having poor heat regulation themselves as a result of poor blood flow in and around the organs. The levels of microwave and/or infrared radiation, emitted from VDUs, however, are not high enough to cause such effects. Measurements of infrared radiation in the environment of a working VDU are less than 1 per cent of outdoor and indoor levels and microwave levels are one thousandth to one billionth the level of the strictest safety regulations (Grandjean, 1987).

Low-frequency electromagnetic fields

The radio frequency band is divided into bands ranging from ultra high frequency (UHF) to extremely low frequency (ELF). The high-frequency range can produce heating effects similar to the microwave and infrared frequencies, but only in excessive quantities. As yet, little is known about the effects of very low and ELF radiation.

Low-frequency radiation is emitted from the flyback transformer in a VDU. This controls the electron beam, sending it back to the top of the screen, approximately 100 times per second. The result of the activity is the emission of ELF radiation. Manufacturers, however, usually shield these emissions, resulting in only minute quantities being present. Indeed, the level of radio frequency radiation in the vicinity of a VDU is comparable to the ambient levels generated by radio transmitters in urban areas. These levels are some 2,000 times lower than those permitted by the strictest safety regulations (Grandjean, 1987).

The concern about low-frequency pulsed magnetic fields was generated by a group of Spanish researchers who reported that chick embryos had been deformed by exposure to such fields. Faults were found with the researchers' methodology, however, and similar experiments have not produced the same results (Mild, 1986). Bergqvist (1986) argues that there is no indication that the low-frequency electric fields have any disabling effect on the foetus. Nevertheless, the debate persists.

Electrostatic fields in combination with light air ions

An electrostatic field is caused when ions flow from a negatively charged body to a positively charged body or vice versa. In the case of the VDU and its operator, the VDU is positively charged – from between zero to 10,000 volts (Jeffreys, 1984) – and the human body is also charged, this time negatively (up to over 1,000 volts in some cases). The electrostatic fields which may occur in front of VDU screens may vary according to different makes and the distance of the operator from the screen (the strength of the field diminishing as the distance increases). A majority of measurements have shown electrostatic fields around VDUs in the order of 0 to 20 kilovolts per metre (kV/m), with a figure of 2.4 kV/m at normal working distance. The combination of an operator and a VDU can generate fields of between 0 and 15 kV/m (Grandjean, 1987).

Apart from an occasional mild electric shock, the presence of static electricity at these levels is harmless. Concern has been expressed, however, about the effect of static on ions in the air, causing a decrease and/or an imbalance. This, in turn, has been suggested as causing stress reactions which may interfere with pregnancy. What impact does exposure to excessive positive ions, ion deprivation and increased deposition of airborne particles have? Jeffreys (1984) refers to two studies: one, in 1972, examined stress and irritability among air traffic controllers and pointed to the problems of an imbalance in atmospheric ions; the second, in 1976, anticipated a number of medical symptoms which would be experienced by one-third of the population if negative ions were withdrawn from the atmosphere. These included: irritability and anxiety, headaches, neck/back pain and asthma, apathy, depression, nausea, fatigue, allergies and anorexia. These symptoms were found to occur more frequently among women than men and among older, rather than younger, people. However, Bergqvist (1986) argues that there are no facts to support the assumption that static causes an imbalance in ion concentrations or that any such change can have adverse effects.

Ultrasound

The claim that ultrasound emission from VDUs could cause adverse pregnancy outcomes is not supported by any facts according to Bergqvist (1986).

Polychlorinated biphenyls (PCBs)

PCBs are used as an insulating fluid in some VDUs, and the chemicals have been linked to the occurrence of abnormal foetuses in seabirds and sea mammals. It has been suggested by one study that VDUs leak such chemicals, but this has been contradicted by other studies (Bergqvist, 1986).

Other factors

It should, perhaps, be noted that many of the arguments listed above are kept alive by the media. This may be due to the newsworthiness of a possible link between VDUs adverse reproductive outcomes; more likely, but less newsworthy, causes are stress and worry, and sedentary work.

Stress and worry

Stress may occur in an automated workplace for a number of reasons (see 'The causes of stress' p.51), in addition to which may be added the fear of adverse pregnancy outcomes. The possible link between stress and spontaneous abortion was first mooted some time ago and apparently remains unsolved. To compound the issue, as stress increases so, too, does smoking and drinking, each with known harmful effects on foetuses.

Sedentary work

MacKay (1984) suggests the possibility that sedentary work may adversely affect the pregnancy outcome, although this has (to the authors' knowledge) not yet been confirmed or repudiated.

What can be done to minimize the risk to pregnant VDU workers?

It is obvious from the above discussion that there is insufficient evidence to suggest that the VDU is a source of harm to the unborn child. As soon as one likely cause is repudiated, another enters the debate. Currently, the issues which remain unsolved are: stress, sedentary work and low-frequency pulsed magnetic fields. Until they are resolved, managers should take all precautions to minimize any possible risk to pregnant VDU workers.

By law, the employer has no special duty for his employees during pregnancy (given the current state of knowledge), nor does he have any special duty to protect the foetus. If, under the Congenital Disabilities, Civil Liability Act 1976, he were to have special duty, the employer would possibly be liable to a child born defective because of his negligence.

Thus, although, at present, this is not a problem because no proof exists that employees are at risk, if the climate of opinion changes, employers could become liable (Suter, 1985). However, lack of proof is no justification for lack of sympathy. The case of the dismissed library assistant, Mrs Hazel Johnstone, hit the headlines in both the regional newspapers and nationally in the library press. The Highland Regional Council dismissed Mrs Johnstone for refusing to operate a VDU while she was pregnant. An industrial tribunal was critical of the unsympathetic handling of the case by the Council and ruled it as unfair dismissal. It was considered unfair, because the journal literature contained (and still does) conflicting reports on the alleged health risks. Suter (1985) makes three points concerning this case. The first is that there were other staff available to take over Mrs Johnstone's duties during her pregnancy, the second, that she operated the VDU for only two hours per day, thus making it relatively easy to find alternative duties, and third, that, as with all tribunal rulings, no precedent can be taken from this case. A second case, however, in which Miss Lesley Bradley was awarded £8,000 compensation by an industrial tribunal ('VDU award', 1986), might indicate that, despite the absence of precedents, tribunals look sympathetically on such cases.

It is suggested, therefore, that employers should, where possible, be sympathetic to employees' fears, although it is recognized that not all situations lend themselves easily to finding alternative work for VDU operators. Nevertheless, it is desirable that a pregnant employee should, if they wish, be moved to suitable alternative employment during the course of their pregnancy, with no loss of pay or seniority. Indeed, laws to this effect have been passed in Canada and Ireland, and unions, such as the Council of Civil Service Unions, are arguing for it to become law in the UK. However, the employer should not take all the responsibility. Braidwood (1988, p.35) cites Ross Edwards, an electronic engineer studying the biological effects of electromagnetic radiation, who argues that there is little point in 'coming off a VDU at work unless you are also prepared to review your home environment, because so much there – the radio, the TV and the microwave, for instance – could also be having an effect'.

7 OTHER ALLEGED HEALTH HAZARDS

Cataracts

Another risk to health alleged to be caused by VDUs is cataracts. Cataracts are opacities which occur in the lens: they can cloud the vision and, in some cases even cause blindness.

Cataracts may occur spontaneously, especially in people above the age of 60, although it is said that approximately 25 per cent of people have lens opacities which do not affect vision (Grandjean, 1987). Thus the disease is fairly common and by no means always a threat to sight. Indeed, cataracts which impair vision are rare, affecting only approximately 7 per cent of the population.

It is considered likely that 20–30 per cent of VDU operators aged between 20 and 50 and 40–50 per cent of those over 50 will in the normal course of events develop cataracts but, of these, only a small proportion will suffer visual defects.

The causes of cataracts

The causes of most cataracts, especially in the younger age groups, are not well known. They may be present at birth or may form in early childhood, linked perhaps with the occurrence of an intra-uterine infection or genetic metabolic deformities. Later in life, the causes may be due to trauma, metabolic and degenerative disorders or exposure to an agent such as high steroid intake, previous eye surgery or radiation (Hamdy, 1987).

Ionizing radiation can also cause cataracts (Scalet, 1987), as can exposure to either ultraviolet, infrared or microwave radiation at sufficiently high levels. These types of cataracts, which form as the result of radiation, can usually be distinguished from spontaneously occurring opacities. Zaret (1984) cites a paper by Sir Stewart Duke-Elder which indicates that a range of rays from ELF (Extremely Low Frequency) to soft ionizing frequency radiation could cause cataracts. Zaret himself believes that radiation emission from the VDU is a cause of cataracts.

Despite Zaret's claim, the causal connection has not been substantiated by any other author. Zaret's findings are well below the anticipated number of operators who would be expected to have cataracts if 25 per cent of the population suffer them from natural causes.

The cataracts that Zaret examined appeared to have been caused by radiation, but, as the radiation emissions from VDUs are well below those of the surrounding environment, it would seem unlikely that the VDU would be the cause in all but the most extreme of cases and some other source of radiation may be the cause. Grandjean (1987, p.201) puts it more strongly: 'There is therefore not the slightest reason to suspect that VDUs have a cateractogenic effect.'

Facial rashes

There have been various reports in the literature of people becoming 'sunburned' at work or experiencing a facial rash whilst working on the VDU. The rashes usually disappeared a few hours after finishing work. After much research, a probable link has been found between the electrostatic field caused by VDUs and their surroundings and the appearance of facial rashes.

The causes of facial rashes

As stated in Chapter 6, an electrostatic field may surround the operator and his VDU. As a result, it is suggested, there is a strong attraction of negative ions in the atmosphere towards the VDU screen and, at the same time, a propelling of positive ions towards the operator's face, and then out of the immediate vicinity. Thus, initially, there is an increase in the positive ion concentration inhaled by the operator, and also in minute airborne particles which hit the operator's face at a faster than normal pace. The transfer of ions can be further accelerated if surrounding objects are also highly charged – for example, synthetic carpeting.

Jeffreys (1984) reports a Norwegian study which demonstrates that the deposition of particles on the skin increases by a factor of more than 10 in the presence of electromagnetic fields and that this increase is mainly in submicron elements. Because sulphur and chlorine are concentrated in the submicron section of the atmosphere, the skin is being bombarded, not only with particles flying at a greater than normal speed, but also with concentrations of these particular chemical deposits.

The effect of an alteration in the ion balance is still being investigated, but 'the belief that electrostatic phenomena may cause facial skin rashes of VDT [VDU] operators [is now accepted] to be a reasonably well-justified hypothesis' (Grandjean, 1987, p.200).

What can be done to prevent them?

It is only in a minority of cases that skin rashes occur. Should a member of staff be affected, there are a few things that can be done to help reduce the static in the vicinity of the VDU and the operator:

- The problem may be in the type of carpeting, so a change of carpet to a low static one can help, although it is very costly. A cheaper alternative may be to use a rubberized antistatic mat. These can be purchased in a range of shapes and sizes and are designed to fit under the chair and the desk. Prices are in the region of £100 for a 4′×6′ mat. Smaller mats, costing around £30, are available for desktop use, fitting under the VDU and keyboard. It may also be possible to persuade the cleaners to spray the carpet with a static-reducing chemical or even with fabric conditioner which has static-reducing properties.

- Low air humidity can also encourage a build up of static. This can be rectified in various ways (see Chapter 12, 'Relative humidity', p. 128).
- One operator found that using a skin moisturizer two or three times per day was helpful. This may have helped to act as a barrier against the submicron particles.
- Antistatic sprays for use on the display screen can be used, but this can become expensive over a long period.
- Earthed screens which are placed in front of the VDU screen can reduce static and glare. These cost in the region of £50.
- It may be helpful to rearrange the workstation so that the operator is not directly facing the VDU (Rycroft and Calnan, 1984); in some instances, improved workstation design may occur by positioning the VDU screen slightly to one side (see Chapter 13).
- Another cause of static may be the operator. Clothes, particularly underwear, are often made from synthetic fabrics which can encourage a build-up of static. A change in the type of clothes or underwear worn may, therefore, cause less static. It is left to the reader's discretion to decide how to phrase this delicately to his staff!

Static electric shocks

The presence of a field of static electricity around the VDU and the operator can also result in mild electric shocks being received when the operator comes into contact with an earthing device such as the computer or a metal filing cabinet. While not dangerous, such an experience can be unpleasant and unnerving. The methods of reducing static listed above, if implemented, will reduce or eliminate the occurrence of static electric shocks.

Photosensitive epilepsy

Epilepsy is an unfortunate illness which, thankfully, is experienced by only a small proportion (0.5 per cent) of the population (Newmark and Penry, 1979). Of these, only between 1 and 3 per cent are subject to photosensitive epilepsy – that which is triggered by a flickering light or a particular kind of pattern.

The use of VDUs at work, with their pattern of lines, is a possible trigger for such an attack. An estimated 50 per cent of photosensitive epileptics, however, experience their first seizure whilst watching television (Cakir *et al.*, 1980). This event is likely to have occurred in the 10–14 year age group which is typical for the onset of the illness. A person is therefore likely to be aware of their illness prior to starting work.

Anyone who is unsure should seek medical advice, and known photosensitive epileptics should avoid VDU work. The Employers Medical Advisory Service (see Appendix) can offer help to photosensitive epileptics.

Legionnaires' disease

This disease is carried in air conditioning systems typical of those used in many libraries and for mini/mainframe computers. Although it is an unlikely risk, it may be advisable to have the systems checked periodically. Symptoms are similar to those of flu and may therefore go undetected for some time. Less serious diseases, such as humidifier fever, may also be caused by the air conditioning system, causing headaches and fevers.

Cabling

Loose cabling can be hazardous and cause accidents. It should be hidden to avoid tangling with the feet of the operator or of passing staff and users.

Fire

Any electrical equipment could short-circuit and start a fire, and a computer is no exception. Fire extinguishers (carbon dioxide or powder variety – not foam) or fire blankets should be near at hand and staff should know how to use them. Large computer installations may be situated in a single, easily sealed room, in which case a fire prevention system, such as halon gas, can be used. Adjacent rooms should not contain hazardous materials such as volatile chemicals (Brownrigg, 1982).

Laser printers

Laser printers have also been rumoured to be a health hazard, reports in the literature claiming that the ozone released can cause breathing problems and blindness but, as yet, there is no evidence to support this.

Summary and recommendations

Much space has been given to the discussion of possible health hazards arising from working with VDUs because there are many underlying fears and uncertainties concerning them. If the possible hazards and the solutions are more clearly understood, efforts to reassure staff may be more effective. It is hoped that some of the preceding discussion will offer a basis from which further research may be undertaken. In this field, as in many others, discoveries are continually being made and it is necessary to keep abreast of developments.

It must be remembered that many positive effects will also result from automation. Jobs may be enhanced, reducing stress and some health risks may even be eliminated! The 3-part issue system, for example, often used carbonless paper which has been reported to cause headaches, eye, skin and respiratory irritation (Hedge, 1987). It is thus important to recognize that automation is not necessarily the bringer of health problems but that it, like many other aspects of modern life, has its own problems which need to be recognized in order to be overcome.

Many of the complaints, especially the postural problems, have arisen from intensive VDU work. At the present time, much library use of VDUs is not intensive, with the obvious exceptions of data input and word processing. Other uses tend to be interactive and offer frequent breaks. Nevertheless, unless steps are taken to prevent them, health problems do still occur, and will do with increasing frequency, as automated systems become more common. It is equally important, however, that in reacting to the problems, matters are not made worse by encouraging a change which is inappropriate or by causing unnecessary alarm and concern (Scalet, 1987). For example, some have argued that pregnant operators should automatically be transferred from VDU work. This may mean moving an issue desk clerk to photocopying or shelving, jobs which require a great deal of standing and possibly lifting of heavy items; both tasks pose a greater known risk to the unborn child than the VDU. Similarly, if action is taken to shield the operator or the VDU, the logical conclusion will be that there is something which needs shielding. Consequently fears may be aroused, rather than laid to rest.

In conclusion, a variety of recommendations may be made to help reduce the real or imagined health problems.

- Recognize that managers have a legal obligation to take 'reasonable care' for the safety of their employees.
- Listen to staff complaints and fears sympathetically and do what is required to boost morale or reduce fear, but remember that complaints about fear of health problems may mask other, deeper anxieties.
- Develop good workstation design and ensure that staff know how to rearrange the furniture to suit themselves.
- Ensure good job design.

Part III WORKPLACE DESIGN: WORKSTATION CONSIDERATIONS

8 BASIC CONSIDERATIONS

Introduction

Library staff are one of the most expensive and complex components of a working library system, and it is therefore important that they are given an environment which maximizes their ability to work. This means, in effect, ensuring that the workplace is designed to maximize their comfort, safety and well-being. As a result, productivity is also likely to increase. Studies reported by Mason (1984), for example, show that 'ergonomically designed' furniture can increase productivity by 10–25 per cent.

Throughout the discussion, the terms 'workplace' and 'workstation' will be used. 'Workplace' refers to the complete working environment including the heating, lighting and decoration, while 'workstation' is used to describe the more limited concept of the furniture and other components which together make a unit for one or more operators to perform a particular task.

In addition to increasing productivity, good workplace design can reduce, or even eliminate, many of the health complaints listed in part two. This, in turn, can have positive effects on productivity. Spilling *et al.* (1986) found that substantial savings were made when the workstations of manufacturing operators were redesigned. This was due to a significant improvement in worker health, which in turn resulted in reduced labour turnover and sick leave. Reduced turnover brought about a consequent reduction in the costs associated with recruiting and training new staff.

The importance of ergonomics has been filtering through the library literature but, sadly, does not seem to have penetrated many libraries. Two reasons for this are suggested by Pinder and Storey (1986): 'a plethora of old buildings and a lack of new money'. Certainly, it is true that the initial outlay on the computer system is difficult for many libraries to obtain and, faced with the recurrent costs of maintaining the system, the idea of extra expenditure on compatible furniture is often rejected.

It is not just finance which causes ergonomics to be overlooked. Danks (1984, p. 14) reported the findings of a study by the ergonomist at

Butler and Cox who found that 'ergonomics was very much a secondary issue when compared to the problems, particularly of first-time users, of getting data processing applications to work'. There is also, it seems, a belief among some managers that people are infinitely adaptable and able to adjust to work with a new system; with other managers there may simply be a lack of awareness of what to do to provide an ergonomically sound environment.

Whatever the reasons, many examples of poor workstation design can be found throughout the library and information world. Busch (1984, p.7) performed a study of information workers in New England and found that:

> . . . while most respondents used adjustable seating, only 12% had their own chair for VDT [VDU] use. None had adjustable desks. Only 6% could adjust the screen angle, or used copy holders. 50% cleaned their screen less frequently than once a month. Only 29% could control the ambient lighting in their workspace. And only 12% had an anti-glare filter installed on their VDT screen.

Nevertheless, it makes good economic and managerial sense to give thought to workplace design and the following chapters outline the factors which need to be considered. This chapter discusses some of the basic considerations which need to be examined such as the users of the workstation, the tasks to be undertaken, and posture.

Users of the workstation

Ideally, the workstation should be designed to suit the person who works there. This is relatively straightforward with a single-user workstation but, where several people share a common work area, it is more difficult. People vary in their physical characteristics (see Chapter 1) and in their opinion about what is comfortable. Slovak and Trevers (1988) found, for example, that people such as designers, programmers and scientists who used the VDU for 10–50 per cent of their working day could tolerate ergonomic and postural problems more easily than could secretaries and accounts personnel who spent 80–100 per cent of their time performing VDU work. They suggested two possible reasons for this. In part, the increased tolerance may have been due to the relatively small amount of time spent at the VDU, but it may also have been affected by the attitude of the staff towards their job. It was suggested that those with more creative work, such as designers and programmers, had higher job satisfaction and used the computer in a creative rather than routine fashion. The happier people were with their working life, the more likely they were to tolerate poor workstation design and its attendant problems. The more dissatisfied they were overall, the less happy they would be with inadequately designed workstations. Because people vary in size, shape and attitude, shared workstations can create particular problems. A shared workstation should therefore be adjustable, especially if a range of tasks is to be performed.

The task(s) to be performed

Libraries vary in size from 'one-man bands' to large academic institutions and thus the tasks performed are equally varied, ranging from one-man operation to a large organization where job specialization is the norm. The effect of certain tasks on workstation design is considered further in Part V.

Sitting posture

Incorrect posture, as outlined in Chapter 3, is one of the factors which will induce backache, headaches, pains in the neck and shoulders, digestive and circulatory problems. Sitting has several advantages (Grandjean, 1980):

- It takes the weight off the legs.
- It helps to avoid unnatural body postures.
- It reduces energy consumption.
- It places fewer demands on the blood system.

Nevertheless, it brings with it a set of problems:

- It causes a slackening of the abdominal muscles ('sedentary tummy').
- It leads to curvature of the spine which in turn is bad for the digestive organs and lungs.

The most severe problem, however, is for the back. The spine is formed by separate bones (vertebrae) which are linked together. Between each vertebra and the next is a kind of cushion or disc. These discs can gradually become worn away or flattened, or the viscous fluid, which causes the cushioning effect, may be squeezed out. This, in turn, can cause the tissues and nerves to be trapped or strained and can lead to various pelvic disorders, and lumbago. Bad seating can accelerate deterioration of the discs by increasing the pressure on them. The following relative pressures have been recorded (Grandjean, 1980):

Posture	Disc Pressure (%)
Standing erect	100
Lying flat on the back	24
Sitting, trunk erect	140
Sitting, bent forwards	190

which indicate that the pressure on the discs is greater when sitting than when standing.

The ideal position for the back is one which allows the spine to adopt a gentle 'S' shape, which is the norm when we are standing. What happens when we sit slumped is that the small of the back curves outwards rather than inwards thus increasing the pressure on the discs. However, in order to sit up straight, muscular effort is involved, which can be tiring over long periods. Thus the ideal posture is one which enables the back to retain its 'S' shape, but does not require muscular tension to retain it.

The pressure on the discs is also increased if weights are being lifted. Cataloguing, acquisitions and issue desk work require frequent lifting of piles of books. This can compound the pressure on discs if an incorrect posture is adopted. Yates and Karwowski (1987) found that men could lift greater weights when standing than when sitting and, whilst seated, could lift a greater weight in a forward position than when twisting the trunk through 90° to lift something from the side. Therefore care needs to be exercised when designing a workstation to ensure that a combination of simultaneous lifting and twisting is avoided.

It is not just lifting or twisting which affects the pressure on the discs, however. The sitting posture itself can have an effect, and there is much debate in the literature on what is deemed to be the ideal posture. There are currently three main schools of thought:

- those who favour the 'traditional' upright posture, with hips and knees at right angles;
- those who argue for a forward-tilting chair, extending the angle between the spine and the thigh;
- those who prefer a backward-leaning posture.

Upright posture

The traditional view of the VDU worker's posture is of an erect back, horizontal thighs, and knees at right angles. The elbows should be bent at 90° and the forearms should run parallel to the thighs, a short distance above the work surface. However, studies have shown (Tougas and Nordin, 1987) that only 10–20 per cent of people sit in the recommended upright position.

Forward-tilting posture

Mandal, reported by Pheasant (1986), argues that the seat should not be horizontal, but should slope forward. This means that the top part of the body can lean forward to perform tasks, such as keyboarding or writing, whilst, at the same time, the downward slope of the thighs retains the natural curve of the spine and prevents cramping of the internal organs. Unfortunately, the downward slope of the seat can cause the body to slide, enforcing the operator to hold himself still by pushing with his feet. If the chair is on castors, this pushing action can cause the chair to move away from the desk (see Figure 8.1a).

An alternative is the kneeling (or 'Balans') chair, which has a forward tilting seat and a kneeler on which to rest the knees and take the weight of the body (see Figure 8.1b). Anecdotal evidence, however, indicates a split of opinion between those who find the kneeling chair more comfort-

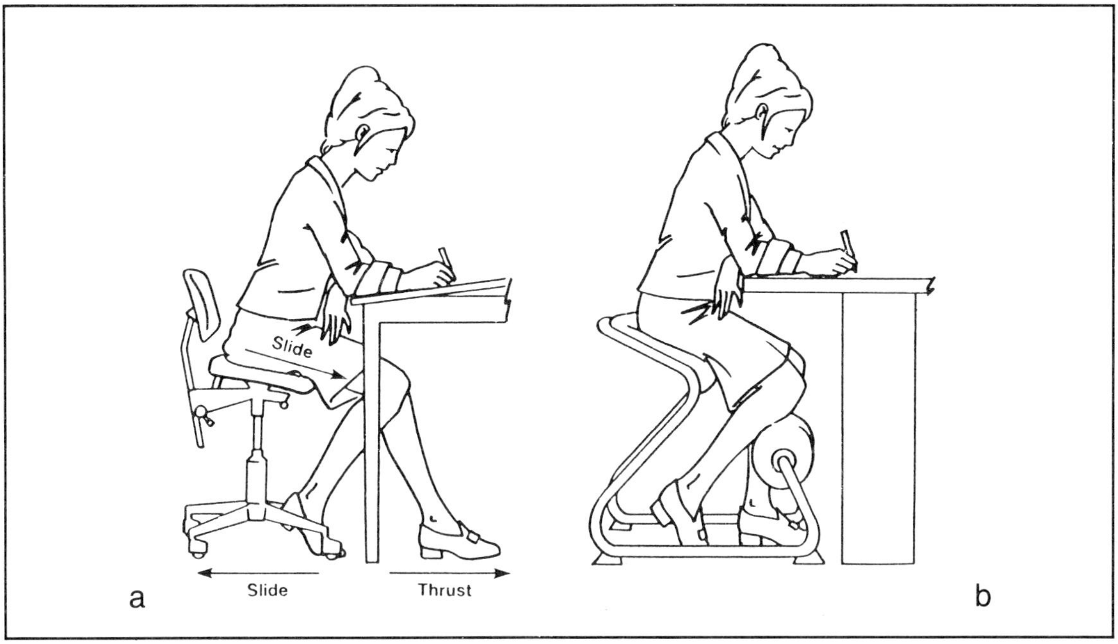

Figure 8.1 New approaches to seating: The forward-tilting seat (left) and the kneeling chair (right)
Source: Pheasant, 1986, p. 163

able than the traditional office chair and those who find that, after a period, their leg muscles tire and their bodies begin to slump. Indeed, Tougas and Nordin (1987, p. 209) concluded that the main disadvantages of the kneeling chair were: 'the absence of lumbar [lower back] support to rest back muscles and limit slumping, and the pressure on the knees which can produce discomfort'. In addition, Brunswic, as reported by Pheasant (1986, p. 164), found that:

- When seat angle and knee angle were independently varied, forward tilted seat positions did not result in a lumbar posture which was significantly different from that obtained with a horizontal seat and the knees at right angles.
- The lumbar posture of subjects using a kneeling chair in writing and typing tasks was not significantly better than when they used a conventional office chair.

He concluded that 'the kneeling chair does not offer any measurable advantage over well designed and correctly adjusted chairs of the conventional type'.

Backward-leaning posture

Research has shown that a backward-leaning posture not only reduces the disc pressure, but is one in which the back muscles are more relaxed. An angle of between 110° to 130° between the seat and the back has been found to be both comfortable and relaxing. Certainly, the backward-leaning posture is one which is adopted by several operators, even though the chair is not designed to support them.

It has also been found that a lumbar pad – that is, a support in the small of the back – can help to reduce disc pressure. However, when this pad is used in combination with a backward-sloping angle of between 110° and 120°, disc pressure is lower than when standing (Grandjean, 1987). However, Mandal (1986) argues that the backward-sloping position requires greater bending of the neck which could cause pain in the neck over time.

What is the ideal posture?

The traditional upright posture has been selected as the basis for the workstation designs which follow, although some reference is made to the backward-leaning posture. This is not intended as a recommendation of either posture but is to form a working base until the experts reach a conclusion. This cannot, however, be assumed to be 'ideal' (Cakir *et al.*, 1980, p. 153):

> . . . no matter how well the ergonomic criteria for VDT [VDU] workplace design are satisfied in practice this, in itself, is no guarantee against postural discomfort. If the job itself requires the individual to maintain a static position or permits only a few essential movements, some degree of postural discomfort is still to be expected.

In other words, to design a workplace which forces an individual to maintain a so-called optimal posture *cannot* be optimal.

Fleischer *et al.* (1987) in a study looking simply at the weight distribution of 16 seated subjects during an experiment concluded that, even in this limited study where operators performed identical tasks, the sitting behaviour of each of the operators was different. Given a range of ages and tasks, it is even more likely that operators will exhibit a range of preferred postures: leaning back, crossed legs, one foot tucked under the body, bent forward with feet crossed behind and so on. Thus, the optimal workplace is affected by two factors: the need to be able to adjust the workstation to suit all the staff who will use it and the need for those operators to express their individual preferences in sitting behaviour and still have freedom of movement.

9 WORKSTATION COMPONENTS

In order to attain a comfortable and efficient workstation, detailed consideration needs to be given to the components which make up the workstation and to how these may be assembled. The following items together form a workstation for VDU work: chair(s) desk(s); VDU; keyboard; barcode reader; mouse and/or touchscreen; printer; and document holder. For details of copy or document holders see Chapter 13; the other items are described below.

Chairs

The chairs used by the operators should be able to support the body in a position which minimizes disc pressure and muscular tension. For the purposes of discussion, the upright posture with the knee and hip joints at right angles has been chosen. This position may be attained using the conventional typist's chair, although high-backrest and tilting chairs are now available (Figure 9.1).

Figure 9.1 The traditional typist's chair and a newer high-backed chair (with optional tilting mechanism)

Investment in new office furniture may be required, as few libraries already have such chairs. Indeed, in some, even traditional typist's chairs are not available.

Grandjean (1987) performed studies which office workers using the traditional typist's chair and a high-backrest chair. The latter was favoured by 75 per cent of the group studied, and 89 per cent of the group favoured a high-backrest chair which tilted backwards.

Recommended features for chairs

There are a number of recommended features for chairs. These include the following:

- Chairs must be able to cater for forward, upright and backward-leaning positions. This means that the backrest must be adjustable in angle. It must also be possible to lock the chair at the required angle to ensure support in the forward-leaning and upright positions.
- The backrest should be 48–50 cm in height (between the surface of the chair and the top of the backrest) and between 32 and 36 cm wide. Ideally, it should have a lumbar pad at a height of 10–20 cm above the lowest point of the seat surface, and should be concave towards the top of the backrest to accommodate the outward curve of the spine by the thorax. The whole backrest should be concave in the vertical plane, and should be adjustable forwards and backwards to ensure that equal support is given to all sizes of users.
- The seat should measure 40–45 cm across and 38–42 cm from front to back. If it is too small, it may cause discomfort and restricted circulation. The seat should be slightly hollow and raised a little towards the front to prevent the operator sliding forward. The front should curve away from the legs (a waterfall edge) to prevent cutting in on the thighs and restricting circulation.
- The back and seat should be upholstered to approximately 2 cm thickness and covered with a non-slip, permeable fabric.
- The chair should be adjustable in height to cater for different operators and it should be easily adjusted from the sitting position. A gas or hydraulic mechanism is probably better for this purpose. However, it has been reported (Ralli, 1988) that in libraries where gas-lift chairs are used operators do not always adjust them, so some training is required. Footrests should be provided for short operators to take the weight of the lower leg and to reduce pressure on the underside of the thighs. Footrests should be adjustable in height and angle.
- Chairs should have a stable (five-point) base. It should not be assumed, however, that five points will necessarily guarantee stability, especially when sitting on the edge or leaning backwards. Grandjean (1980), in a study of 261 male and 117 female office workers, found that they spent 15 per cent of the time sitting on the edge of their chairs and 42 per cent of the time leaning backwards. Stability, therefore, needs to be ensured, and this can only really be

done by testing the furniture prior to purchase.

- Armrests can help to reduce the pressure on vertebral discs but, if poorly designed, can impair operator performance. It must be possible to move the arms comfortably to perform the required task, to be able to move the chair as far under the desk as required and to get in and out of the chair easily.
- Casters help the operator to move the chair easily, but can cause the chair to become unnecessarily mobile during intensive keyboarding.
- A swivel seat is preferred to prevent the operator twisting unnecessarily. It may be helpful if the swivel can be locked in position for stability when needed.
- High chairs or stools may be needed on the issue desk to bring the operator to the same level as the standing customer. With high stools, a footrest is essential both for stability and to prevent restricted circulation in the thighs. If no footrest is provided, the operator will find his own, which may in turn result in poor sitting posture.

Unfortunately, developments in backrests for operators' and typists' chairs have not been echoed in many high stools. Frequently, those which are advertised are meanly upholstered in plastic or are bare wood with a narrow seat and backrest, neither of which are adjustable. An exception is the draughtsman's chair offered by some furniture suppliers. These are, in essence, tall typists' chairs with a built-in footrest.

Many librarians argue that the cost of new chairs is prohibitive. With prices ranging from £125 to over £400 per chair, it seems unfortunate that, given an expenditure of several thousands of pounds on automated equipment, more could not be found to provide for operator comfort. It even may be cost-effective to purchase appropriate chairs in terms of increased productivity. Paznik (1986, p. 20) reports a study undertaken by T. J. Springer, the president of an ergonomics consulting firm: 'In the study of 120 people, it took 15% less time to accomplish the same amount of work in an adjustable setting as it did on standard desks.' Given the cost of a data entry operator as £5,000 per annum, an increase of 15 per cent productivity over one year would be the equivalent of £750 worth of work. For that price, three *good* chairs could be purchased.

Although the library manager may spend less time at the terminal, his increased productivity could result in even greater savings because of his higher income. Thus a manager should be able to justify the acquisition of a chair befitting his status (whatever that may be!).

Desks

A desk is an important aspect of the working environment, providing both storage and a surface on which to work. In choosing a desk, several factors need to be considered:

- Does it provide adequate storage for the task to be performed and the operator(s) who will use it?

- Is the storage space appropriate (for example, does it incorporate hanging files, shallow drawers, deep drawers) for the task to be performed, and is it easily accessible from a sitting position? Some tasks – for example, online searching – require easy access to heavy manuals. It should be possible to reach these without twisting, bending or stretching.
- Is there room for the operator to move his feet? Some workstations have spars or trailing wires which restrict movement. Lack of movement can cause cramped muscles and aching joints and can restrict the posture of the upper body.

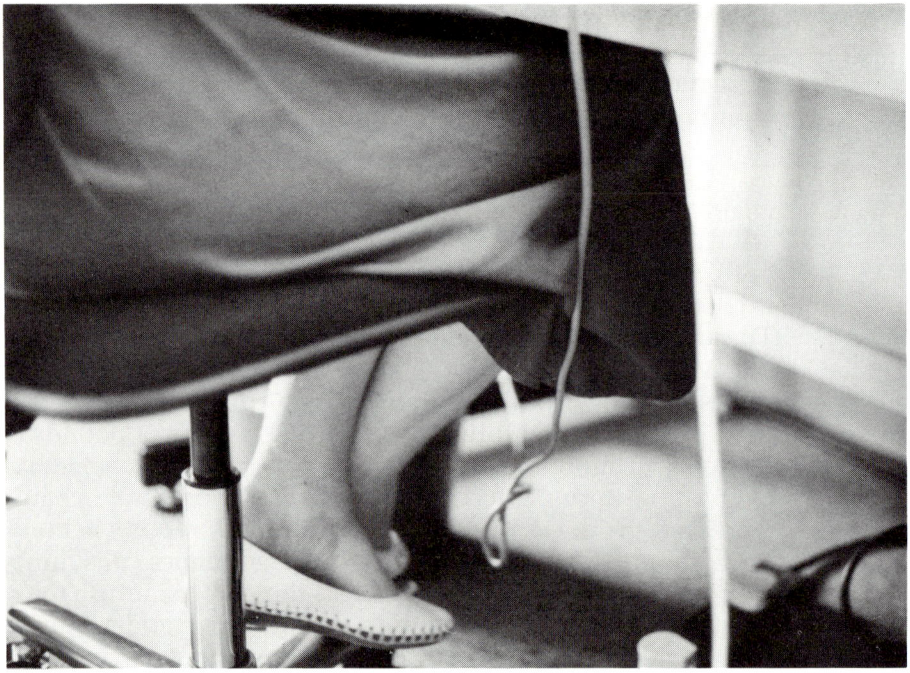

Figure 9.2 Trailing wires can restrict foot movement Photo by Rick Dowling

- Is there adequate thigh clearance? Desks with drawers immediately below the working surface often do not provide adequate thigh clearance. The recommended thickness for the worksurface is 2.5 cm. Much thicker and thigh clearance may be impaired, much thinner and the desk may not be strong enough to support the weight of an operator leaning to write and a computer or terminal.
- Is the desk surface of a matt finish rather than shiny or white, to cut down reflected glare?
- Is the desk surface sufficiently large to accommodate all the items necessary for a task *and* a VDU. In libraries, in particular, workstations

tend to become cluttered, and it is unrealistic to expect to add a VDU to the clutter without creating additional workspace (see Figure 9.3).

- Is the desk the correct height for keyboarding? According to researchers, the 'home row' of keys (ASDF etc. on the QWERTY keyboard) should be level with the elbow. The visual distance is also an important consideration. If the desk is too far away from the operator to see comfortably, he will be forced to bend down to bring the source document or keyboard into focus. On the other hand, too high a desk will cause the operator to hunch his shoulders, putting strain on the neck and shoulder muscles, or put the elbows out to the sides, causing stress on the muscles in the upper back.

  If both keyboarding and writing are performed at the workstation, it may be desirable to invest in a split-level desk which enables the keyboard to be situated at a lower level so that the optimum heights for both writing and keyboarding are obtained.

  Adjustable desks are also available, but they are expensive. The use of such desks ensures that the individual preferences of operators may be met. In practice, however, it is unlikely that they will be adjusted by each operator, particularly as some types can be difficult to adjust. The main purpose of having adjustable desks would be to ensure that a workstation could be designed to suit a single operator exactly, rather than anticipating that people will meticulously readjust the furniture every time a change of shift occurs.
- Special desks may need to be considered, for example to house Online Public Access Catalogue (OPAC) terminals. The Gaylord Caroussel, which houses six terminals in a hexagonal arrangement is one possibility. According to Gaylord, the Caroussel uses 15 square feet less than conventional workstations. Wires are tunnelled through the unit to a 10-outlet power source. The work centre measures 25 inches high, 78 inches wide and 70 inches deep (35 inches for a half-unit housing three workstations). Each individual workstation is 39 inches wide.

Recommended dimensions for workstation chair and desk

No chair or desk can be considered in isolation as it is possible to have a well designed chair, but one which is too low for the desk and vice versa. Thus it is the relative dimensions of both the chair and the desk which need to be considered (see Figure 9.4).

It is recomended that:

- The distance between the floor and the seat surface should be 38–54 cm, although experiments cited by Grandjean (1987) indicate that a range of 32–55 cm is preferred by some operators. The chair should be height-adjustable.
- The distance between the seat surface and the desk top should be between 27–30 cm. If it is greater, the operator will have to raise his shoulders or push his elbows to one side.

Figure 9.3 Library workstations tend to become cluttered photo by Rick Dowling

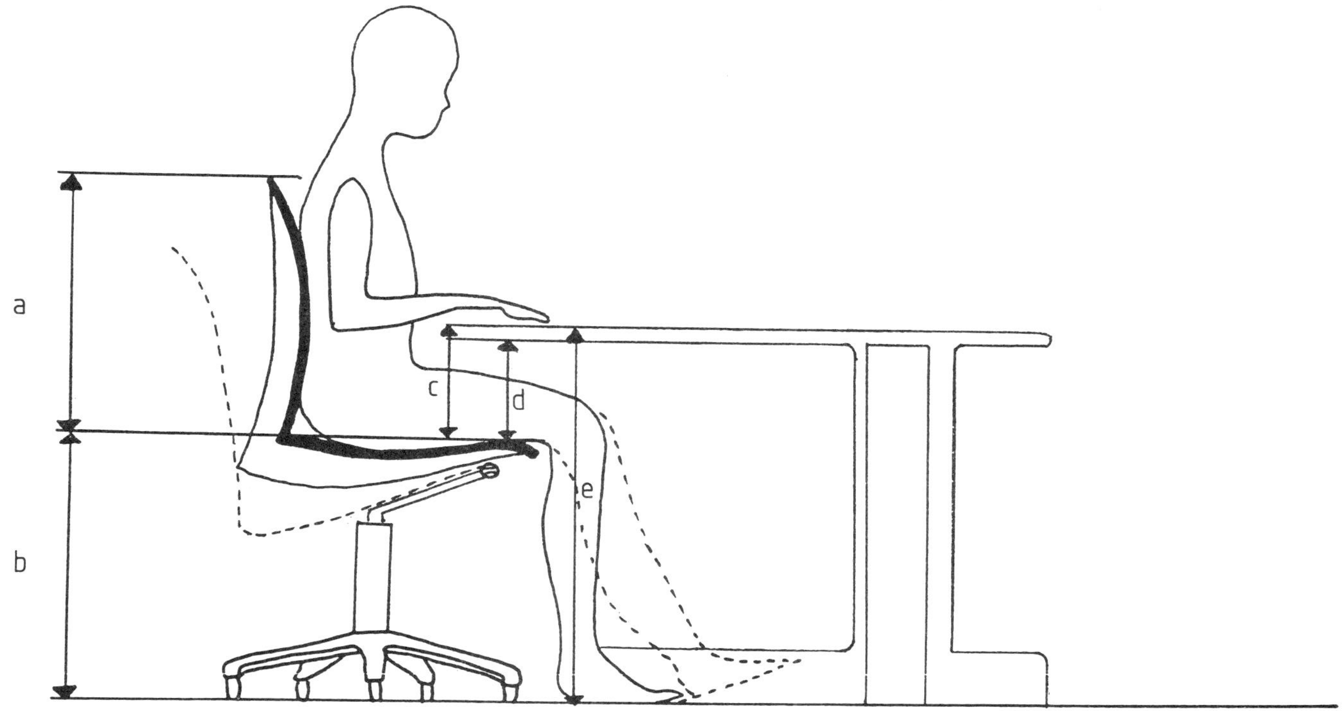

Figure 9.4 To show desk and chair dimensions

- The distance between the surface of the seat and the underside of the desk should be at least 17 cm. It may have to be greater for men, or to cater for heavy outdoor clothing worn by library users in winter (add 35 mm for the latter). This is to ensure adequate thigh clearance. The following heights are recommended by Apex (1985):

Worksurface of desk	Fixed for females	Fixed for males	Adjustable
for writing	69 cm	71 cm	67–79 cm
for keyboarding	61 cm	64 cm	52–67 cm

The level of the desk surface for keyboarding is lower than for writing to allow for the thickness of the keyboard from base to 'home' row (ASDF . . .). Ideally, the 'home' row should be level with the operator's elbow, although Grandjean (1987) cites research which indicates that a large number of operators prefer keyboard heights above the recommended level. Mandal (1986), however, argues that higher desks (87 cm) encourage better posture, although no reference is made to the effect of these heights on the neck and shoulder muscles.

Not all adaptation to improve workstation design has to be costly. Sometimes it is necessary only to undertake a minor adjustment, such as ensuring the desk housing the microcomputer is lower than the one used for writing.

The above recommendations apply mostly to European and North American men and women. In Table 1.1 (p.20), however, the anthropometric measurements could be seen to vary with different groups of people, such as the Asian and Far Eastern races. In libraries employing large numbers of such staff, it may be desirable to use the figures in Table 1.1 and to adjust some of the furniture to suit the relatively small stature of these people. However, as a rule, it is best to cater for the 95th percentile and to assist smaller operators by the use of adjustable chairs and footrests to bring them to the required height.

Public library users offer the widest range of body sizes to be catered for, ranging from children to old people. The needs of disabled people should also be considered as desks intended for standing adults will be too high for wheelchair users and may prove tiring for users with other kinds of disability. Figure 9.5 illustrates a public user terminal designed with disabled users in mind.

Space also needs to be provided for users' personal belongings. Problems can occur when OPAC terminals or microfiche catalogues are housed in individual study carrels (1,000 mm wide), as insufficient space is available for personal belongings such as the ring-binders or lever-arch files typically used by students. It is preferable to place two terminals on a flat desk or double carrel thus offering greater flexibility.

The use of chairs with castors can aid access to public terminals, but it is appreciated that this may not always be desirable, either from the point of view of cost, or because of misuse by playful library users.

Figure 9.5 *A public access terminal designed with disabled users in mind.* Photo courtesy of Carolyn Householder, Nashville State Technical Institute, Tennessee

The VDU

The VDU itself, and features such as its shape and colour, can affect operator comfort and well-being. Many features of the VDU screen and its display – flicker, character stability, character contrast, reflections, screen colour, positive versus negative images and the character set have already been discussed in Chapter 4. This section will deal with aspects of the box which houses the cathode ray tube.

Ferris (1987) lists a number of standards for the casework design and finish of a VDU, issued by various national and international bodies. Some of these are included in the list below:

- It must be possible to tilt the VDU screen to adjust its angle, and a range of between −5° to +20° from the vertical is suggested. It is also desirable that the operator should easily be able to adjust the angle by simply pushing or twisting the VDU casing.
- It should be possible to swivel the screen easily from one side to

another. Some screens, such as the IBM PC range, are not easily moved, and in such cases it is possible to mount the screen on a turntable which allows the screen to be swivelled and, often, tilted to the required angle (see Figure 9.6).

Figure 9.6 A tilt and turn VDU support. Photo courtesy of Acco Europe Ltd

- It may be desirable to alter the height of the screen, and a minimum range of adjustment of 110 mm is recommended.
- The outer surface of the VDU (and keyboard) casing should be a light colour and matt or semi-matt to minimize reflected glare.
- The colours for indicator lights should be:

  red – for danger
  yellow – for caution
  green – for safe operation

Thus the red indicator light used on most disk drives to show that the disk is being accessed, contravenes this standard (although it could be argued that it is a warning that the disk is being accessed and any interference could cause damage of the disk and/or the data). Red lights which are used to show that the shift lock or caps lock keys are in operation are also inappropriate. Green would be preferred.
- Major controls, such as on/off, brightness and contrast should be clearly labelled and within reach of the seated operator.

The keyboard

The keyboard should be separate from the VDU to enable the operator to choose his own preferred position relative to the VDU screen. This may vary according to the type of tasks performed (see Chapter 13, 'The tasks performed', p.134). The keyboard should be lightweight and easy to move around on the worksurface, but with grips underneath to prevent unwanted movement during keying.

Other factors, such as keyboard height, slope, layout and geometry, key design and the use of special purpose keys, and audible feedback can also affect operator performance levels and comfort. These are discussed below.

Keyboard height

The height of the keyboard is a debatable issue. Apex (1985) recommends that the keyboarding desk height should be 61 cm for women and 64 cm for men. The thickness of the keyboard, however, will determine the height of the 'home' row of keys, which is the important issue. The recommended keyboard thickness is 30 mm (Ferris, 1987), but many keyboards, particularly older models, are thicker. Grandjean (1987) cites a study that indicates an acceptable range of home row to floor heights of between 73 and 85 cm. Thus it may be possible, even with a thicker keyboard, to adapt the workstation to provide an acceptable keyboard height. However, with thicker keyboards it may be necessary to provide wrist support.

Other factors compound the problem of keyboard height. These are rather eloquently outlined by Pheasant (1986, p. 193):

> Assuming that the correlation between sitting elbow height and thigh thickness is zero [i.e. that there is no link between the two, ed.], there is a space of 80 mm between the elbows and the tops of the thighs in women and 85 mm in men. The home row of a typical typewriter stands some 70 mm above the desk top, which is itself some 25 mm in thickness. The inescapable conclusion is that the recommended typing posture is a physical impossibility for around half the population – irrespective of the design of the desk. The surprising narrowness of the gap between the elbows and the thighs argues strongly in favour of low profile keyboards and strongly against drawers or other obstructions in the kneehole of the desk.

Keyboard slope

The slope of the keyboard should be gentle, and an angle of 15° or less is recommended. The slope of the keyboard can often be adjusted by 'legs' underneath the back of the keyboard. Usually, however, these enable only two options – with or without the legs and, in some cases, the use of the legs can obstruct access to the disk drives if the keyboard is placed directly in front of the drives.

Keyboard layout

In Chapter 3, it was reported that rapid repetitive work could cause repetitive strain injuries (RSI). The layout of the alphanumeric keys may be to blame, according to Professor Ferguson of Sydney University (Hobday, 1985). There are various layouts available which are discussed below. The most common and, according to Professor Ferguson the cause of RSI, is the QWERTY keyboard.

QWERTY

The QWERTY keyboard (so-called because the top row of keys spells the word 'QWERTY') is also known as the Sholes layout after its inventors), the Sholes brothers (see Figure 9.7). QWERTY was developed in 1873 and has now become the most popular in English-speaking countries. French-speaking countries, however, still prefer AZERTY (Stobie, 1986).

```
Q W E R T Y U I O P
 A S D F G H J K L ;
  Z X C V B N M , .
     space-bar
```

Figure 9.7 The Sholes (QWERTY) keyboard

Originally, keyboards were alphabetical but it was found that keys began to jam. The Sholes brothers therefore designed a keyboard which separated those letters which frequently co-occur in the English language, such as 't' and 'h' and 'q' and 'u'. Their aim was reliability in keyboard operation, not speed. Nowadays, however, typewriters are not prone to jamming and, of course, neither are computers (at least not in that way!). It is therefore technically possible for the keyboard layout to be changed. This may be desirable as there are a number of problems with the QWERTY layout (Gilad and Pollatschek, 1986):

- Forty-eight per cent of motions are one-handed, that is, it is necessary to move one hand to touch two consecutive keys. It is possible to type more quickly if keys are pressed by alternate hands in the same way that it is possible to beat a drum faster with two hands than with one. It has been estimated that one-handed motions take 75 per cent more time than two-handed actions.
- It is necessary to move to a row other than the 'home' row (ASDF . . .)

for 68 per cent of key strikes; 70 per cent of strokes should be on the 'home' row.

- The hardest motions in typewriting are one-handed motions to one or two rows above the 'home' row. Some writers are of the opinion that the number of these motions forced by the QWERTY layout is excessive.
- Fifty-six per cent of all strokes are performed by the left hand. For most people, it is desirable to have the majority of strokes undertaken by the right hand.
- The QWERTY keyboard induces awkward, one-handed fingering and inequitable loading on each finger. Consecutive strokes, ideally, should be with different fingers and, preferably, with different hands.

Because of the problems of the QWERTY keyboard, several alternatives have been designed, each with its own group of supporters.

Dvorak layout

In 1932, August Dvorak, Professor of Statistics at the University of Washington, developed the following layout (Figure 9.8):

?	,	.	P	Y	F	G	C	R	L
A	O	E	U	I	D	H	T	N	S
'	Q	J	K	X	B	M	W	V	Z
				space-bar					

Figure 9.8 The Dvorak keyboard

The aim of this design was to improve the efficiency of the typist. All the vowels are placed on the 'home' keys of the left hand and the most-used consonants on the right hand set of 'home' keys. Thus more than two-thirds of the key strokes are on the home row (Stobie, 1986). The remaining keys are laid out according to their frequency of use, with the more common on the top row and the less common on the bottom. The loading on the little fingers has also been reduced.

Dvorak himself maintained that typists trained on this keyboard were faster and more accurate than those on QWERTY, and one world speed record is held by a Dvorak typist, Barbara Blackburn, at 170 words per minute (wpm). QWERTY records of 176 wpm and 216 wpm feature in the Guinness Book of Records, however (Stobie, 1986).

The Dvorak layout has become popular in the USA and some computers even offer a software selected Dvorak layout (for example, the Apple IIc), but this is not available on European machines.

Maltron

Lilian Malt has given her name to the keyboard she devised in the 1970s (Figure 9.9):

1	2	3	4	5	6	7	8	9	0
B	W	F	L	M	Y	G	U	P	C
A	N	E	H	R	D	O	T	I	S
Z	X	?	Q	;	.	:	V	J	K

Figure 9.9 The Maltron keyboard

The vowels have been split between the right and left hands of the home row and the more commonly used letters are under the first and second fingers. Frequently occurring letter combinations, such as th, ion and so on can be typed without the fingers leaving the home row.

The keys are set into a sculptured curving surface and are divided into five main groups. The main alphanumeric keys are divided into two blocks, one for each hand. Each thumb has eight keys to control, set again in a block for each hand. A separate numeric keypad is situated in the centre of the keyboard.

Under each palm, the keyboard casing is raised, offering support for the hands when they are not being used for typing.

Alphabetical layout

It has been suggested that novices might type faster using an alphabetical arrangement. Norman and Fisher (1982) quote other writers who fail to support this view, finding that novice typists typed as rapidly on the QWERTY keyboard as on those which are arranged alphabetically. In Norman and Fisher's own study the operators, on average, achieved a typing speed of 13 wpm on a QWERTY keyboard compared with 7.8 wpm on two alphabetical keyboards and 7.1 wpm on a randomized keyboard layout.

In all these studies, although non-typists were used, the bias in favour of QWERTY may have been due to the slight acquaintance that most adults have with the standard keyboard layout. Children with no experience of the traditional layout may be better subjects for such an experiment.

An alphabetical arrangement may seem logical to those unfamiliar with the QWERTY keyboard. Certainly, it could be argued that it would place less mental strain on the operator as the characters would logically follow one another. Norman and Fisher (1982) used the following keyboard layouts in an effort to assess the effectiveness of alphabetical keyboards.

They concluded that alphabetical keyboards are not always as logical as they might seem. For example, they argue, suppose that one is typing the word 'flower' on the alphabetical-1 keyboard. The 'f' is located on the top row near the middle of the keyboard. Where should one look for the 'l'? Logically, it should follow to the right of the 'f' but, in this case, one must look to the left to find it. Thus the logical arrangement of an alphabetical keyboard cannot be assumed.

The diagonal alphabetical keyboard

A	D	G	J	M	P	S	V	Y
B	E	H	K	N	Q	T	W	Z
C	F	I	L	O	R	U	X	

space bar

The 'alphabetical-1' keyboard

A	B	C	D	E	F	G	H	I	J
K	L	M	N	O	P	Q	R	S	
T	U	V	W	X	Y	Z			

space-bar

The 'alphabetical-2' keyboard

;	.	,	A	B	C	D	E	F	G
H	I	J	K	L	M	N	O	P	Q
R	S	T	U	V	W	X	Y	Z	

space bar

The 'alphabetical-3' keyboard

A	B	C	D	E	P	Q	R	S	T
F	G	H	I	J	U	V	W	X	;
K	L	M	N	O	Y	Z	,	.	

space-bar

The 'alphabetical-4' keyboard

A	B	C	D	E	F	G	H	I	J
K	L	M	N	O	P	Q	R	S	T
,	.	;	U	V	W	X	Y	Z	

space bar

The random keyboard

C	Y	I	F	M	G	Z	D	N	J
Q	O	X	H	B	T	R	W	L	
V	A	U	P	K	E	S			

space-bar

Figure 9.10 Keyboard layouts used by Norman and Fisher (1982)

Kathleen Gardner has devised an alphabetical keyboard (Figure 9.11) which, it is claimed, is so easy to use that eight year-olds can reach speeds of 90 wpm within a few months (Barclay, 1986). It is also considered to be ideal for the 'hard-pressed executive [who] could learn to use this keyboard in half an hour'. The keyboard is designed to be easy to learn because of the logical positioning of the keys relative to one another. The design also takes into account the problem of muscular control of the little fingers, requiring them to press only one key each, A and Z. The Gardner keyboard has been patented in Britain and the USA.

Velotype

Launched in 1983, the procedures of the Velotype claim that it can improve typists' productivity by up to 300 per cent. The high typing speed is obtained because syllables rather than individual letters can be input. The Velotype keyboard is arranged in three sections. Vowels occupy the centre sections and the left- and right-hand sides mirror each other comprising the same set of consonant keys. A high degree of intelligence is built into the keyboard to ensure that certain letters are keyed in the correct order – for example, S*P*OON could not be misspelt *PS*OON, but DRO*PS* would be allowed. Thus, it is possible to press several keys at the same time and for the Velotype to sort them into order and it is the simultaneous action which enables speed of input.

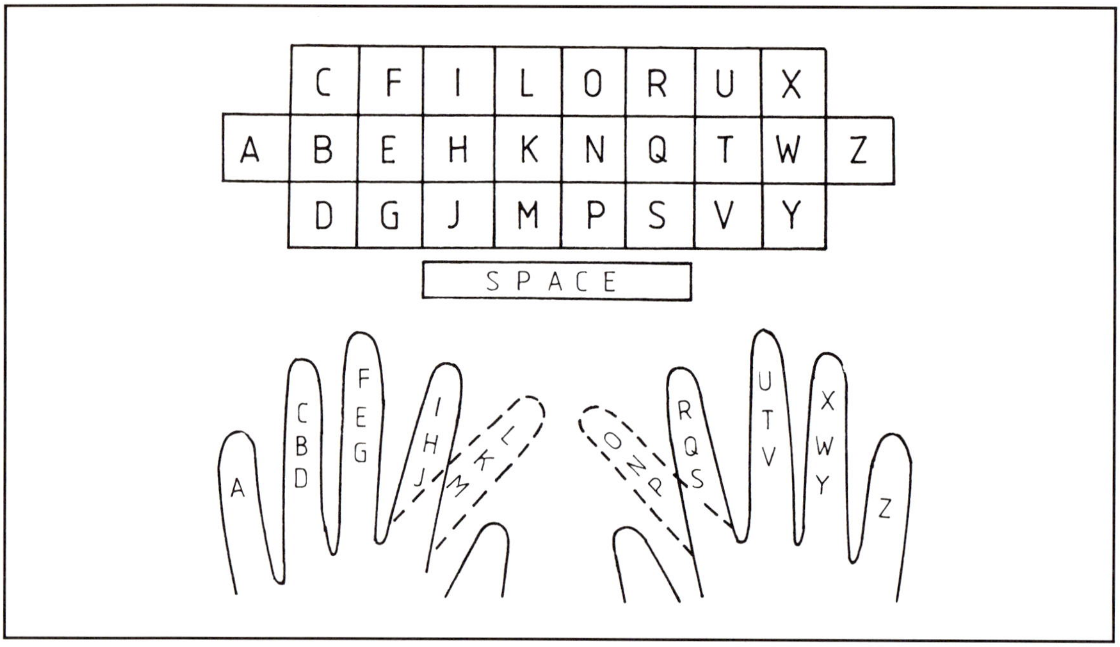

Figure 9.11 Kathleen Gardner's new keyboard layout
Source: Barclay, 1986

It can be concluded that, although there are many advantages of altered keyboard layouts (they can be easier for novices to learn and may offer improved productivity through increased typing speeds) there is one overriding disadvantage, which is that QWERTY is the *de facto* standard. A large number of typists are already trained on the QWERTY layout and, as Norman and Fisher (1982) showed, even non-typists' acquaintance with the QWERTY layout makes them reasonably proficient. Norman and Fisher also concluded that, 'for the expert typist, the layout of keys makes surprisingly little difference'. Similarly, while it is desirable to encourage children to use computers in libraries, there is no evidence to suggest that they are deterred by the QWERTY layout. It is therefore questionable whether they should be encouraged to use a special keyboard type, such as Kathleen Gardner's, in the children's section only to have to 'graduate' to another layout on transfer to use of the library's main stock.

Another problem is that, in many libraries, there are a number of typewriters which are used for miscellaneous items, such as typing British Library interlibrary loan request forms. These keyboards cannot be redefined, and it may be difficult for staff to switch between the two layouts. A small, but important, point is that as a result of being the standard layout, QWERTY keyboards are usually more robust than their rivals (Stobie, 1986).

For libraries wishing to experiment with altered keyboard layout, however, a cheap and easy way of customizing a standard keyboard is explained by Pollatschek and Gilad (1986).

Keyboard geometry

Some writers argue that it is not the layout of the alphanumeric keys which causes RSI, but the shape of the keyboard. Indeed, some researchers believe that altering keyboard layout without amending its shape, or geometry, could increase the risk of RSI. Most traditional keyboard layouts set the alphanumeric keys in parallel rows, causing the operator's wrist to be twisted during typing. The new lighter keyboards (with the possibility of revised key layouts) encourage faster typing. This, in turn, may cause an extra strain to be placed on the wrists and lower arm. To counteract this problem, ergonomically designed keyboards have been developed which allow the arms to be held in a more natural position. One such keyboard is known as STR.

The STR keyboard was developed in Zurich in 1983 under the direction of E. Grandjean (Buesen, 1984). The alphanumeric keys remain in the same relative positions as the standard keyboard but the parallel rows of keys are replaced by two blocks of keys: one for the right hand and one for the left.

In testing, the preferred dimensions for the keyboard were:

Angle between keyboard halves	25°
Lateral tilt	10°
Horizontal tilt	10°
Maximum distance between keyboard halves	10 cm

Thus the STR keyboard endeavours to offer the best of both worlds, one which is shaped to suit the natural angle of the hands but which does not require a typist to be retrained to a new keyboard layout.

An alternative ergonomic keyboard offering similar features is described by Ilg (1987) (see Figure 9.12). Shaffer (1986, p. 119), however, dismisses all the lengthy arguments about keyboard layout and geometry succinctly:

> Arguing from the viewpoint of the skilled typist I suggest that the benefits of rearranging letters may be small and of altering keyboard geometry perhaps negligible. The greater benefits of ergonomic design are to be found in the characteristics of the keys themselves, their resistance to pressure and their surface contour.

Shaffer also argues that with the light touch now offered by computerized keyboards, the risk of RSI is reduced or even eliminated. 'On the contrary', he writes, 'the finger dance demanded by key layout can promote the strength and suppleness of the hands, and it is possibly less tiring to move the hands continuously than to have them hover in a stationary position.' In summary Shackel (1989) reports that three major

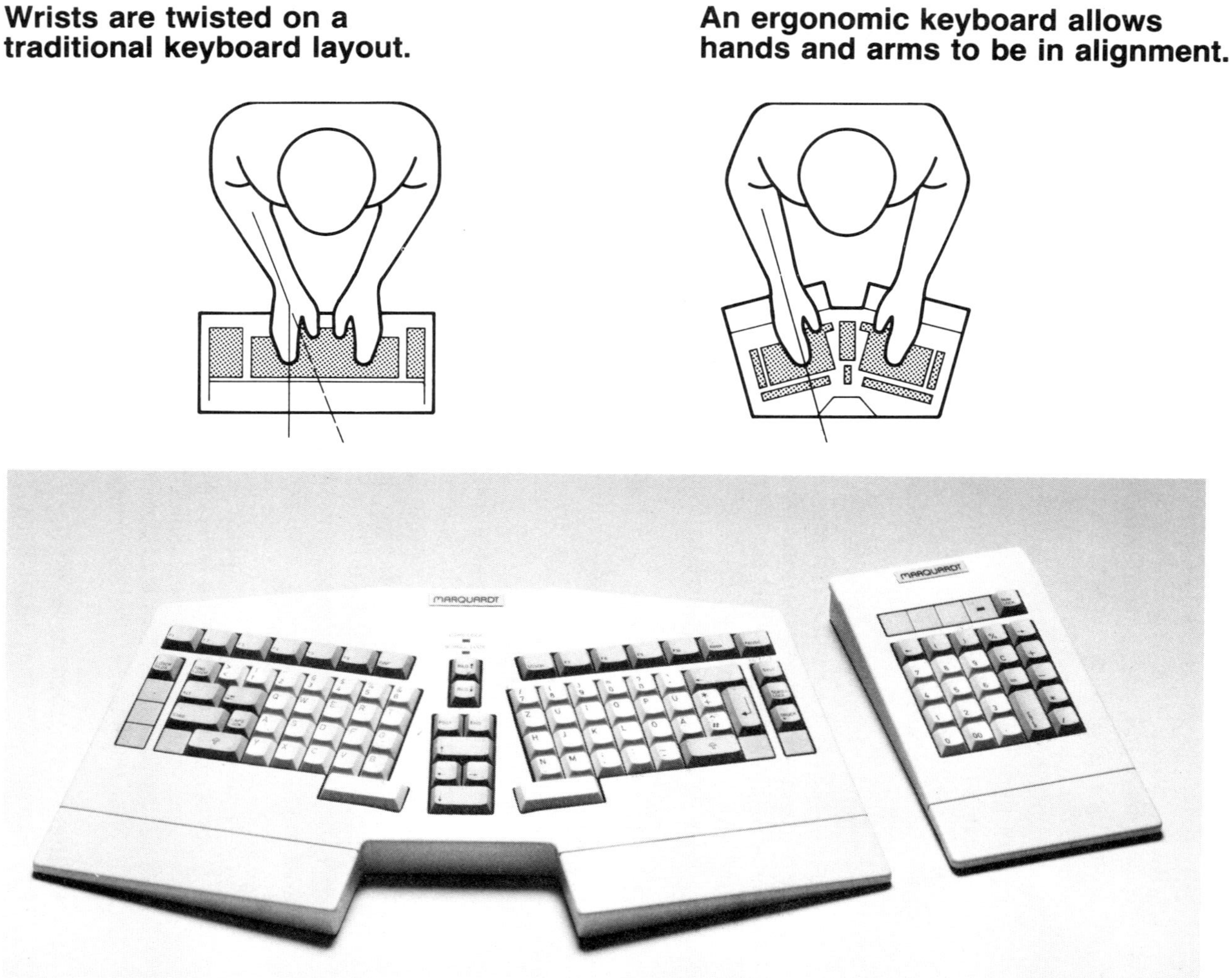

Figure 9.12   *An ergonomic keyboard*. Photo: courtesy of Marquardt GmbH

studies have indicated a change to a split keyboard is to be preferred, although there is little benefit in altering keyboard layout.

The keys

Various factors in key design affect the ease of use of the keyboard.

Key separation

It is recommended that the distance between the keytops should be 17–19 mm (Grandjean, 1987) or 19±1 mm (Ferris, 1987). If keys are too close, keyboarding errors may occur and, if too far apart, the hand is forced to travel too far between keystrokes reducing speed and, possibly, inducing fatigue.

Keytop area

The top of the key should measure 113 mm^2 and be a minimum of 12 mm in any direction (Ferris, 1987). Anyone who is familiar with using the small keys of many of today's pocket calculators will appreciate that a small key size is not synonymous with speed and accuracy.

The top of the keys should be concave as an aid to accuracy. A concave shape reduces the risk of the operator's finger slipping to an adjacent key in error and confirms to the operator in a tactile manner that the finger has located the centre of a key. The minimum radius of the concave area should be approximately 10 mm (Ferris, 1987).

If keys travel only a short distance before coming to a halt, the keyboard can give the impression of being 'hard' and have the effect of jarring the operator's arm. Conversely, if the keyboard is too 'soft' and the key travels a long way before activating the signal, miskeying may occur. Operators may tend to think that the keystroke has been accepted and yet have failed to depress the key sufficiently. The optimum distance for the key to travel is 3–5 mm (Grandjean, 1987). The amount of force required to depress the keys will also affect the 'hardness' or 'softness' of the keyboard.

Audio feedback (usually a click) or tactile feedback (a sensation through the key) may be helpful and aid keyboard accuracy. Grandjean (1987) suggests that the best feedback quality is achieved when the point of acceptance and pressure is located about halfway down the key displacement. If audio feedback (key click) is used in addition to tactile feedback, it is desirable that it can be turned off (if only for the benefit of the person in the same room as the keyboarder!).

The names of the keys should be easily legible under the lighting conditions in which the VDU is being used: a minimum height of 2.6 mm being recommended and a combination of dark letters on a light background preferred (Ferris, 1987). The legends should be resistant to wear and abrasion and should be easy to understand. The word 'ENTER' or 'RETURN' is often more easily understood by novices than the frequently used symbol ↵, and similarly 'CAPS LOCK' is more meaningful than 'TTY CAPS'.

The surface of the keys should be of low reflectivity, as should the legends, to help minimize reflected glare. They should also be dirt-resistant.

Keys should be neutral (beige or grey) in colour rather than black or white (or red, yellow, blue or green) as this is more restful on the eyes. For the same reason, there should be a low contrast between the keys and the keyboard. Some contrast in colour however is useful between different groups of keys – for example, numeric and function keys.

Special purpose keys

In automated library and information work, several keys are needed in addition to the alphanumeric keys. Retrieval from an online catalogue system, for example, may be by ISBN, thus requiring the use of a numeric keypad for ease of input. Function keys, too, play an important role as they help to reduce the memorizing and input of commands. They are used, for example, on some OPACs. The position of these other keys needs therefore to be assessed in order to maximize efficient use of the system. Each type is discussed below:

Numeric keypad

Where numeric data are frequently input, a separate numeric keypad is desirable. Preferably, this pad should be detachable to allow left-handed operators to use it more easily.

The majority of computer and calculator keyboards adhere to the following layout:

7 8 9
4 5 6
1 2 3

If, however, a telephone with a numeric keypad is also frequently used, the operator may begin to experience problems, because telephone keypads are laid out in the following format:

1 2 3
4 5 6
7 8 9

The latter arrangement is usually preferred by novices if the keyboard can be reconfigured.

The BREAK key

The BREAK key can be invaluable as an escape mechanism for the experienced operator, but may have disastrous consequences in some systems if pressed inadvertently. It should be positioned so that it cannot accidentally be pressed. Some systems require the operator to press CONTROL and SHIFT simultaneously with the BREAK key, but while this prevents inadvertent use of the key, the simultaneous use of three keys can cause problems for some users.

Frequently-used keys

Keys, such as RETURN, which are in constant use, need to be positioned so that they are easily accessible. It is helpful if they are larger than the other keys to assist accuracy.

Cursor control keys These should ideally be situated in a logical position relative to one another (see Figure 9.13) to minimize the mental loading and to increase speed and accuracy in their use. On some keyboards, the numeric keypad doubles as a set of cursor control keys and this usually offers a logical arrangement.

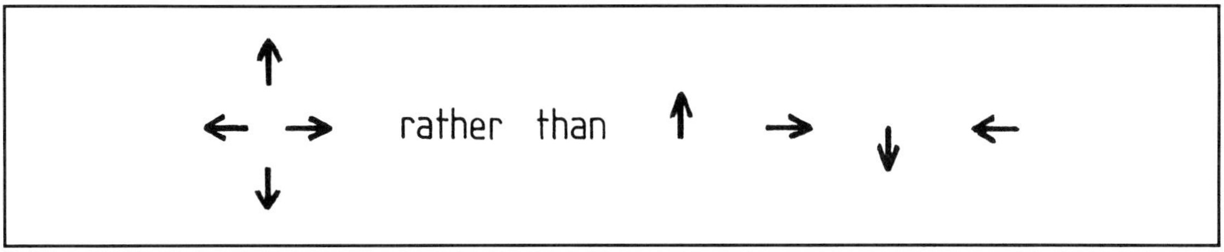

Figure 9.13 To show preferred cursor control layout

Function keys Some keys have functions whose use is specific to particular programs. A set of these should be clearly distinguished by colour and position. Usually they are situated either in a row along the top of the keyboard which correlates well with a prompt line at the foot of the screen, or in a two-column block at the left of the keyboard. The former is often used with a cardboard memory strip which can be placed above the keys, while the latter permits the use of a cardboard template designed to fix round the keys. These are supplied with much commercial software.

Keyboard error signals It is helpful, too, to have an audible warning of keyboard errors, such as when two keys are struck simultaneously (other than on a Velotype keyboard) or when the buffer is full – that is, when the computer is no longer accepting the characters as they are being typed.

Forearm rests Holding the arms in a raised position for typing for any length of time is tiring and this can be compounded by delays in computer response which force the operator's hands to hover unproductively over the keys. It is desirable, therefore, that the operator has something on which to rest his forearms, specially if there are pauses in the keyboarding, such as in a conversational task (see Chapter 13, 'The tasks performed', p.134).

One form of support for the wrists is the surface of the desk. If the keyboard is positioned a little way back from the edge of the desk, it is possible to use the vacated space as a wrist support. However, the thickness and slope of the keyboard should not force the operator to bend his wrists sharply upward nor allow the edge of the desk to cut into the operator's forearm. An alternative is to use a specially designed wrist support which attaches to the keyboard. Other supports have been devised, including forearm and elbow supports. Bendix and Jessen (1986) found, however, that supporting the elbows on armrests mounted on the chair brings about a constrained working posture.

Differences of opinion exist as to what form of forearm support, if any, is appropriate. Grandjean (1987) argues that disc pressure is reduced if the weight of the arms is carried by a support and that less strain is placed on the shoulder muscles if a support is used. In contrast, Bendix and Jessen (1986) found that the loading on the trapezius muscles (those which pull the shoulders back) was significantly greater with wrist support than without and was more marked the higher the support was positioned.

Figure 9.14 The front edge of the desk can be used as a wrist support. Photo by Rick Dowling

In the absence of any recommendation it is best to allow operators to adjust their own keyboard and forearm support. It is suggested, however (Ferris, 1987), that a palm rest should be used with keyboards with home key heights of more than 30 mm.

Barcode readers The main factors to look for in barcode readers are accuracy in reading and ease of manipulation.

Accuracy Frustration can occur if the barcode reader frequently misreads. This may be due to the poor quality of the printed code or the type of code and the way in which the light pen reads. If a barcoding system is to be used it is highly desirable that the accuracy of the reader and the barcodes be tested.

Ease of use

Some light pens read successfully in contact with the barcode while others require to be held a short distance away. It is easier for a novice to use the former type and these should be chosen where possible. Similarly, bidirectional codes (which read in either direction) are easier to operate than those which read in only one direction.

The wire connecting the barcode reader to the computer or terminal also affects ease of use. In one library, the barcode reader was connected to the terminal with a stiff wire which made it difficult to hold the pen at the correct angle; coiled wires with flexible coatings are to be preferred.

In some libraries, laser readers are used to read barcodes but these are still too expensive for the majority of libraries.

Touch screens

Touch screens are used for some library functions – in particular some OPACs. As their name implies, touch screens enable a user to select an item by pointing to it. The simplicity of its operation makes it invaluable as a novice user-interface; because touch screens offer options for the operator to select from, no commands need to be remembered and keyboard skills are not required.

There are disadvantages with touch screen technology, however, not least that the screen requires frequent cleaning! Another problem is that touch screens can only work by offering choices to the user, thus restricting the flexibility available with a command-driven system. In one of the early touch screen OPACs, the user was obliged to select up to 14 different screens before the chosen record was displayed. Thankfully, system designs have since improved.

Use of an OPAC implies that eventually the user will wish to input some data such as author, title or keywords. To do this, a keyboard is required. It is possible to circumvent this problem by displaying a 'keyboard' on the touch screen. The vertical orientation of the screen and its relatively slow speed of response to touch, however, make its use difficult, and for this reason, total dependence on a touch screen is, perhaps, inadvisable.

Mice

Currently little used in the library world, mice are nevertheless increasing in popularity. 'The mouse is, at its simplest, a hand-held device that translates hand movement across a surface into equivalent (scaled) cursor motion on a computer screen' (Whitford, 1986, p. 31). It enables rapid cursor movement, unimpeded by the speed of the key-repeat system or the vertical- and horizontal-only movements of the traditional cursor. It should be of a size and shape which can be held between the thumb and index finger and should be symmetrical for use by right- and left-handed users. The front of the mouse should be lower than the back to accommodate the natural angle of the hand. Any control buttons should be parallel to the fingers to ensure accuracy. The lead should be on the terminal side of the mouse so that it does not become entangled with the hand and the mouse should be easy to move and precise, both in terms of cursor movement and resistance to skidding across the desk surface.

Prolonged use of a mouse, however, has been found to lead to complaints of cramped hands due to gripping the case tightly, stiff fingers from operating the buttons and soreness of the heel of the hand and wrist from rubbing across the work surface (Abernethy and Hodes, 1987).

Printers

Three main characteristics which affect the person at the workplace are noise, print quality and print speed.

Noise

One of the most obvious features of a printer as far as its operator is concerned is the noise it makes when printing. In this respect, printers can be divided into two main types: impact printers, such as daisy wheel and dot matrix printers; and non-impact printers, such as laser, ink-jet and thermal printers. Impact printers, as their name implies, create printed output by 'punching' the shape of the letter through a ribbon onto the paper. This action, needless to say, is noisy. Non-impact printers are quieter, but not silent. Novak (1987) warns users 'never [to] underestimate the noise level of a laser printer'.

The situation of a printer will determine the maximum acceptable noise level. If it is located in the reading room – for use with the OPAC, for example – a quiet printer is a necessity (see 'How to reduce the effects of noise, p.123).

Print quality

The print quality is also important if the presentation of information is to be of a high standard. Nowadays, typesetting via a desk-top publishing system, using a laser printer is the norm for the business world. If the library is to compete, a high-quality printer is a necessary (although expensive) component.

Clarity of print is also important to ensure accuracy. Important information in poor-quality print, or produced by a thermal printer whose output fades in time, may be misread or ignored.

Print speed

Print speed can also be a problem. Too slow a printer can be frustrating for the operator, leading to feelings of underachievement, as well as being inefficient.

Part IV WORKPLACE DESIGN: ENVIRONMENTAL CONSIDERATIONS

10 LIGHTING

Introduction

In order to understand some of the problems concerned with lighting, it is necessary to be aware of the concepts and terms which are used.

The term 'illumination' is used to describe the amount of light which falls on an object or surface. It is measured in units called lux. (For more technical readers, this is 1 lumen per square metre (lm/m^2), a lumen being the unit of luminous flux [Grandjean, 1987].) In practical terms, the measurements may be understood as follows:

50–500 lux – artificial light at night
300–500 lux – recommended office lighting level (Cakir *et al.*, 1980)
500–700 lux – recommended office lighting level (Grandjean, 1987)
100,000 lux – outside in the midday sun on a bright summer's day

Luminance is a measure of light emanating from an object or a surface. Light may have its source in the object – for example, a light bulb or the characters on a (negative) VDU screen – or may be reflected from the surface of the object. Luminance is measured in candela per square metre (cd/m^2).The following examples illustrate the use of this measure in an office with an illumination of 300 lux:

65 watt fluorescent light	10,000 cd/m^2
White paper on table	70–80 cd/m^2
Table surface	40–60 cd/m^2
VDU screen background	5–15 cd/m^2

(Grandjean, 1987)

Several factors need to be considered when discussing lighting. These include illumination levels, light sources, light colour, light glare and other environmental factors, such as the properties of floors, walls and ceilings.

Illumination levels

Too high a level of lighting can cause reflected glare and create dark shadows and marked contrasts between those surfaces which are illuminated and those which are not. On the other hand, if office lighting is too low, although the VDU itself may be readable because of its own luminance, the source documents may be hard to read and visual performance, speed and accuracy may decrease.

Damodaran, Simpson and Wilson (1980) list recommended lighting levels drawn from the Illuminating Engineering Society. These form the basis for Table 10.1.

Table 10.1 Recommended lighting levels for different work areas (adapted from Damodaran et al.*, 1980)*

Task/Area	Recommended Lighting Level (lux)
Storage areas with no continuous work [e.g. stacks]	150
Routine work [e.g. library workroom, library counter]	500
Demanding work [e.g. studying tables of figures, manual literature searching]	750
Fine work [e.g. examining and valuing rare books]	1,000

Note: These levels do not take into account the use of a VDU.

As mentioned in Chapter 4, the eyes must adapt to changing lighting levels. If the source document is well-lit, it will appear very bright in contrast to the screen, and thus the eyes will need to adapt continually between one and the other.

It has been suggested that office illumination should be reduced to a level compatible with the VDU (approximately 200 lux), which is lower than the recommended levels of office lighting. However, in the same way as office tasks vary in visual intensity, so, too, do VDU tasks (for an explanation of the different tasks, see Chapter 13). The following lighting levels are therefore recommended (Table 10.2).

Table 10.2 Recommended lighting levels for different kinds of VDU work

Working conditions	Illumination levels (lux)
Conversational task with well printed source documents	300
Conversational task with reduced readability of source documents	400–500
Data entry tasks	500–700

Source: Grandjean, 1987, p. 40)

The above is one attempt to compromise the need for reduced lighting for the VDU with the need for adequate lighting for the source document. Another is to supplement the lighting with desk lamps to illuminate the source document. These can be very effective, but too great a contrast between the darkness of the screen and the brightness of the source document needs to be avoided. It should be remembered that as a person's age increases, so does both his need for lighting and his susceptibility to glare. The age of the staff needs, therefore, to be taken into account when adjusting lighting levels.

Light sources

In order to illuminate a workplace adequately, it is necessary to make good use of the different sources of light which are available – namely natural light and artificial light.

Natural light

One of the features of the British climate, in particular, is the varying level of sunshine. The weak wintry light gives way in summer to bright sunlight, covered intermittently (and sometimes continuously!) by cloud. Natural light is not a reliable source of illumination and needs to be supplemented by a stabler source of artificial light.

The sun must also be shielded to reduce glare. This may be done effectively using curtains or blinds, or the windows may be coated with a thin film which reflects much of the incoming light allowing only a proportion to pass through the glass and also offers the advantage of allowing staff and users still to see out of the windows. Its main disadvantage is its expense. Net curtains (in a muted shade rather than brilliant white nylon) can be used to good effect and can be left drawn all the time. Lateral or venetian blinds are common in offices, but allow small slits of light to enter, giving a stripey effect. Vertical fabric blinds tend to give a more even, muted shade and are to be preferred. Free-standing screens, or even bookshelves can also be used to cut out glare from windows.

Artificial light

It is necessary to provide a supplement or an alternative to natural light. This may be in the form of overhead room lighting or lighting directed to a particular task. Somerville *et al.* (1986) report their use of an automatic dimmer system which reduced output from the artificial lighting when the light shone into the library. They argue that it maintained constant levels of illumination and saved money. As libraries vary, however, so too do the lighting requirements and librarians would be advised to consider the specific needs of individual rooms and desks.

Room lighting

A workplace may be lit by directional lighting, indirect lighting or a combination of the two.

Direct lighting sends 90 per cent of its light towards its goal. Unfortunately, this form of lighting can create a marked contrast between those areas which are lit and those which fall into shadow. In order to reduce this contrast, it is necessary to ensure that the general level of lighting is sufficiently high. This can create its own problems, as we have already seen.

In the main, directional lighting is created by one of two kinds of bulb. In older libraries, light bulbs will probably be of the electric filament variety. The coil of wire inside the bulb is heated and gives off a warm reddish, yellowy glow, and the result is a pleasant and restful light. Electric filament bulbs can be costly to run, however, and may throw insufficient light and the heat output may cause discomfort and even headaches. In modern buildings, fluorescent tubes are more common. Electricity passes through a gas (usually argon or neon) causing it to glow. These tubes are more efficient than electric filaments, giving a high output and lasting for a long time. They tend to be less troublesome as sources of glare, having low levels of luminance when adequately shielded. A variety of shields are available (see Figure 10.1).

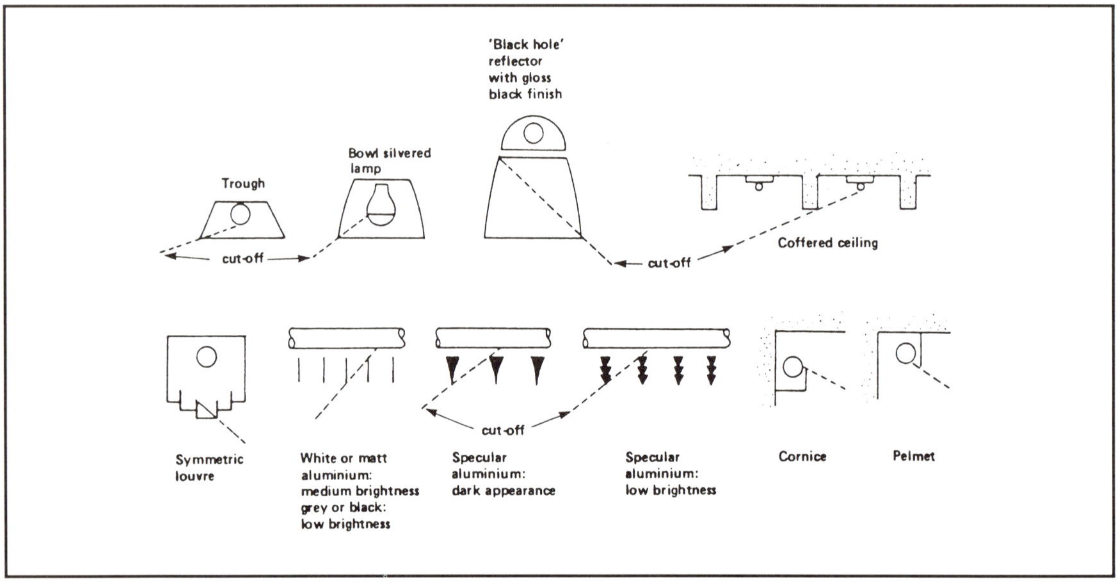

Figure 10.1 Use of reflectors and louvres to control direct glare
Source: Code for Interior Lighting, 1977, Chartered Institution of Building Services Engineers London (superseded by 1984 edition)

The colour of the light varies, depending on the chemical composition of the fluorescent material and can simulate natural light. Flicker can, however, be a problem. Fluorescent tubes emit a flickering rather than a constant light – 100 Hz in Europe, 120 Hz in the USA (Grandjean, 1987). Usually, this is too fast for the human eye to perceive, but research undertaken by Wilkins (1985) found that eye movement was affected by intermittent lighting at 100 Hz, even though it appeared continuous.

One operator interviewed by the authors reported that a particular VDU and fluorescent light combination had caused her headaches. When the lights had been replaced with a different kind, her headaches had ceased. It is possible, therefore, that certain combinations of fluorescent tube flicker and screen refresh rates cause discomfort to some operators. Removing the tubes near the offending VDU and replacing that light with a task light (see below) may be a solution. In Europe, it is now accepted practice to install two or more phase-shifted tubes within one lighting unit to give almost constant light.

Indirect lighting projects most of its light towards suitably coated ceilings and walls which reflect much of this light back into the room. The resulting effect is much closer to that brought about by natural light and is diffuse, causing practically no shadows. However, the bright ceilings and, especially, the walls can produce reflections on the screen and give rise to contrast glare between the VDU and the wall. A number of suppliers now offer vertical lights of some kind: the freestanding pedestal variety; wall-mounted lights (see Figure 10.2); and those which can be affixed to the top of a divider, cupboard or bookshelf.

The lighting colour offered by uplighters varies from a cool blue-green colour to a warmer yellow colour according to the type of bulb which is used. Running costs can be minimized by using low-energy bulbs, and the overall light output can be up to four times greater than fluorescent bulbs (Project Office Furniture, 1984). Economies can also be made by operators switching off individual lights when they leave their desks but, in order to achieve the best effect, uplighters need to be evenly spaced to obtain an even light and, if lights are switched off or moved, the overall effect may be destroyed. In addition, the capital outlay required is not small, either in terms of the new light fittings or the lowering of ceilings and redecorating which may be required (especially in older buildings). It is possible, however, that a combination of reduced energy consumption and increased productivity would repay such an investment in the long term. Hambro Life installed vertical lighting for their data preparation staff and reported a 15 per cent productivity increase as a result. Before installing the lights, the company had tried using blinds at the windows, but this had the effect of cutting out the natural light without reducing VDU glare. After the uplighters were installed, the operators reported fewer headaches (Hunter, 1983), although it may not be the uplighters which caused the improvement. Employees singled out for special treatment have been known to respond favourably even when there was no measurable improvement in conditions – a phenomenon known as the Hawthorne effect.

Direct and indirect lighting can be combined. Uplighters can be obtained with a translucent shade, allowing 40–50 per cent of the light to radiate to the ceilings and walls while the rest is thrown directly downwards; this results in the whole environment being evenly lit.

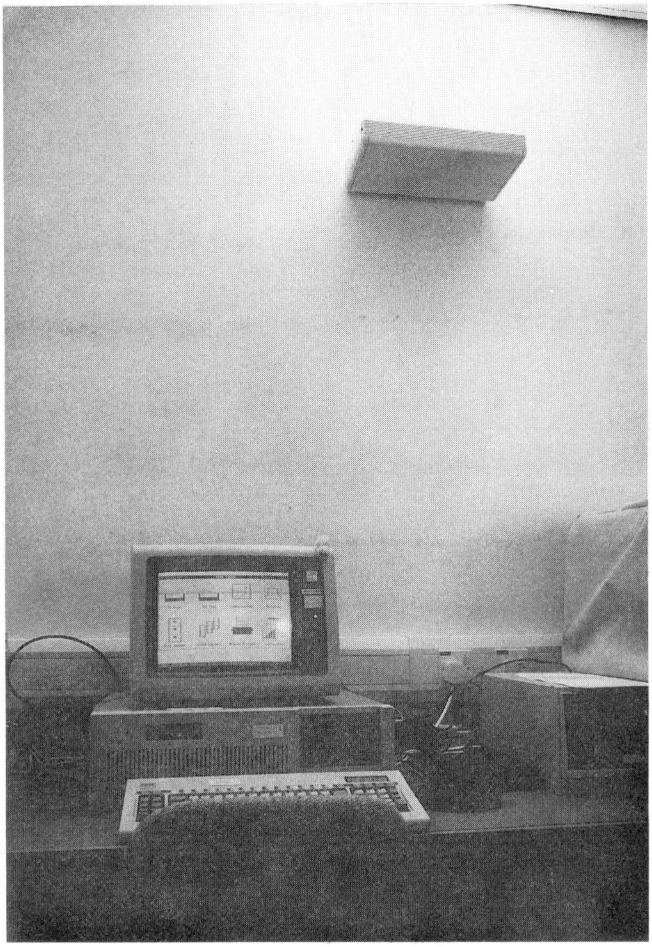

Figure 10.2 A wall mounted uplighter. Photo by Rick Dowling

Desk lighting

To supplement the room lighting for specific purposes, dedicated task lighting can be supplied in the following ways:

Downlighters may be installed below eye-level storage units (see Figure 10.3) or partitions and they can be used to illuminate the required area of the workstations (Hampton, 1985).

Adjustable angle lighting may be purchased. This is available as a freestanding unit or may be fixed to the desk by screws or a clamp fitting. Typical prices for a desk light are in the region of £65–80, but some are available for as little as £34 and as much as £109. Because of the height of the task light, it is necessary to ensure that the bulb is shielded to prevent it becoming a glare source for the operator, and to position it to avoid causing glare to a neighbouring operator. Another problem is that an electric filament bulb gives off heat which may, if poorly positioned, be a source of thermal discomfort to the operator. Cool, more economical and longer-running bulbs are available, but the replacement cost of these is around £6–7 each.

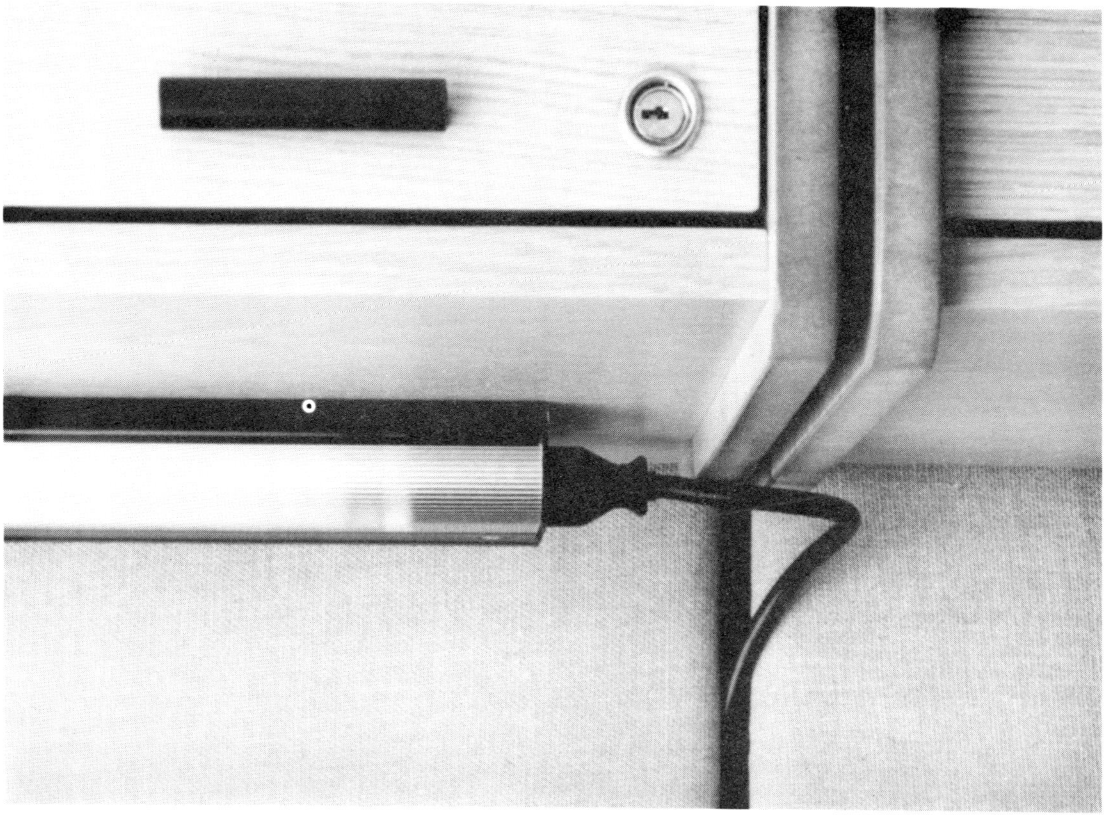

Figure 10.3 A downlighter attached to the underside of a wall-mounted cupboard. Photo courtesy of President Office Furniture

Lighting colour

Whatever form of lighting is used, the colour of the light is important. Light colours have been referred to in the previous discussion and can range from a bright, dazzling white to a warm, cosy yellow or, more precisely from cool, through intermediate to warm. The colour chosen will depend upon the tasks which are to be performed in the room.

A cool light is good for visually intensive work, but can make a room look gloomy, especially if the decor is drab. Intermediate lights are good for clerical work and moderate VDU use, while a warm light is more restful and is suitable for the browsing areas of a library. Problems arise when different types of tasks are performed in the same area, as lamps of different colour appearance should not be mixed in one room (Damodaran *et al.*, 1980). If in doubt, the best solution is to choose an intermediate colour.

Lights as glare sources

As mentioned in Chapter 4, a glare source can be a light fitting or reflected light. The closer the source of glare is to the centre of the visual field, the more performance is impaired: errors increase and productivity is reduced (Grandjean, 1987). The angle of a source of light relative to the operator is therefore important. An overhead light may not be a problem to an operator sitting directly underneath, but may be directly in the line of sight of an operator at the far end of the room, or reflected in a distant screen (Figure 10.4).

Figure 10.4 Lights at one end of a long room can fall directly into the visual field of an operator at the other. Photo by Rick Dowling

Reflected light bounces in much the same way as a tennis ball – that is, in an equal and opposite direction – and so it is important to position the VDU so that reflections do not bounce into the operator's field of vision. The best way to ensure this is to site the VDUs so that the operator's line of sight is at right angles to light sources – both windows and artificial lights. The effect of this is illustrated in Figure 10.5.

However, as Damodaran *et al.* (1980, p. 98) point out, this 'may, in practice, result in too uniform a workroom layout. So, provided that the arrangement of the lamps minimises the risk of glare, these recommendations need not be rigidly adhered to.' Even if the recommendations are adopted, other factors, such as a long window stretching behind the operator, can result in reflected glare (see Figure 10.6). It is also recognized that because of the arrangement of lights and windows in

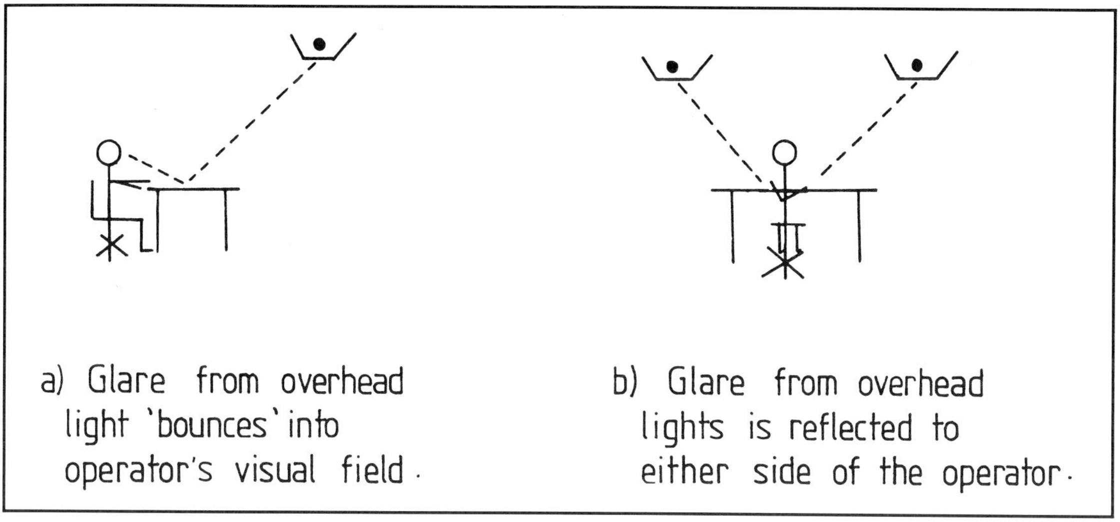

*Figure 10.5 To show the effect of the relative position of light sources (*adapted from Grandjean, 1987*)*

many existing library buildings, it is not possible to achieve the ideal. It may be necessary, therefore, to use one of the devices listed on pages 39–42 to help reduce reflected glare.

Figure 10.6 Long windows, although parallel to the line of sight can cause reflected glare

Other recommendations to help reduce glare are as follows:

- Ensure that there are no highly reflective surfaces in the operator's line of sight (it should be noted that even the protected covers of books awaiting processing may be sources of reflected glare, and need to be positioned out of the line of sight).

- Provide all lights with shades or glare shields.
- As a rule, it is better to use several lamps of lower power than a few, high powered lights to increase the overall level of lighting and to reduce excessive contrasts and heavy shadows.
- Glare can be relative and the perceived glare from a bright light may be reduced by painting the surrounding ceiling a bright yellow or white.

Floors, walls and ceilings

Light is reflected from the surfaces of the working environment. The amount of light which is reflected, the size and colour of the glare source and its distance from the operator will all play their part. The reflective properties of a surface will depend largely upon its finish, shiny surfaces reflecting considerably more light than matt finishes. In order to reduce reflected glare, matt surfaces are to be preferred.

The colour of the surrounding environment will also affect the overall illumination and the emotions people feel towards the working environment. Grandjean (1987, p. 42) lists the following qualities:

- Blue and green seem further away, rooms appear larger, slightly colder and more restful.
- Red and orange seem close, are warm and rather exciting.
- Brown seems very close, rooms appear small, and many people find brown colours to be cosy.

High contrasts between colours should be avoided and, for the most part, restful colours, such as soft pastels prove most popular. It is advisable to select colours of similar brightness for the different surfaces and to avoid eyecatching effects. White ceilings are recommended. Flooring should have very low reflective properties, carpets being good from the point of view of sound reduction as well as light absorption. Strong desk colours should be avoided where possible, but it is appreciated that, in heavily used areas, light colours are impractical.

Paznik (1986) describes the dramatic effects of colour on one company which had darkened the walls in order to reduce glare. Unfortunately, the darkened walls also reduced productivity and energy levels; people were inhibited from communicating with each other and became more irritable. Once the operators were moved to a lighter environment, productivity increased. Mental capacity can be consumed in having to cater for an unpleasant environment. If the environment is improved, therefore, spare capacity is made available for work and, hence, productivity increases.

Recommendations

In most libraries, technological change has overtaken architectural change. For example, in one old library, in order to install automation at all, it was necessary first to install a ring main. The lighting systems in some libraries may be equally archaic. For those librarians with access to sufficient funds to rectify the situation, the following steps should be taken (Damodaran *et al.*, 1980).

- Review the existing lighting scheme, position and type of lighting, position and type of plug sockets and cable runs, position and size of windows, and ask the staff their opinions on the lighting for their work.
- Identify the activities which will take place and where.
- Establish the recommended lighting levels for the different tasks (see Tables 10.1 and 10.2).
- Measure the existing lighting levels using a photometer.
- Consider how to adjust the existing lighting levels to the recommended lighting levels.
- Assess any glare problems, whether direct or reflected, and consider how to eliminate them.
- Install the new lighting system and review it by repeating the above steps after a period of approximately one to two months.

11 NOISE

Introduction

Noise is simply a sound which the hearer considers to be unpleasant or disturbing. Psychologically and physically, people differ in their reactions and receptivity to, and perceptions of, sound.

Sound is caused by a fluctuation in air pressure. The frequency at which the pressure fluctuates is measured in cycles per second; the higher the frequency the higher the pitch of the sound. The loudness of the sound is a function of the size of the air pressure fluctuation and this is measured in decibels (dB). Pitch and loudness together affect the apparent 'loudness' to the human ear; the higher the pitch of a sound, the louder it appears to be. A logarithmic scale has been devised to show the relative loudness to the hearer, using a weighted measure, dB(A) (see Figure 11.1). The scale is logarithmic because, as the sound levels increase, an even greater increase in sound is required for a noise to appear louder:

> 1 dB is the smallest change the ear can distinguish; a 6 dB increase is a doubling of the sound pressure level, although a 10 dB increase is required to make it sound twice as loud (Grandjean, 1987).

Noises seldom occur singly, and should not be considered in isolation. In fact, noise is cumulative. Take, for example, the difference in noise level between a single typewriter and an office containing three or four working typists.

It is not only the level of noise which affects us. Grandjean (1987) lists a number of annoyance factors which form the basis of the following list.

- Noise level.
- The nature of the noise – unexpected and irregular noise being more disturbing than regular noise, and high-pitched noise more annoying than low-pitched noise.
- The nature of the work of the hearer – if a person 'approves' of the source of the noise (for example, if it emanates from the machine he is operating) it will be less annoying than if it is from an outside source.

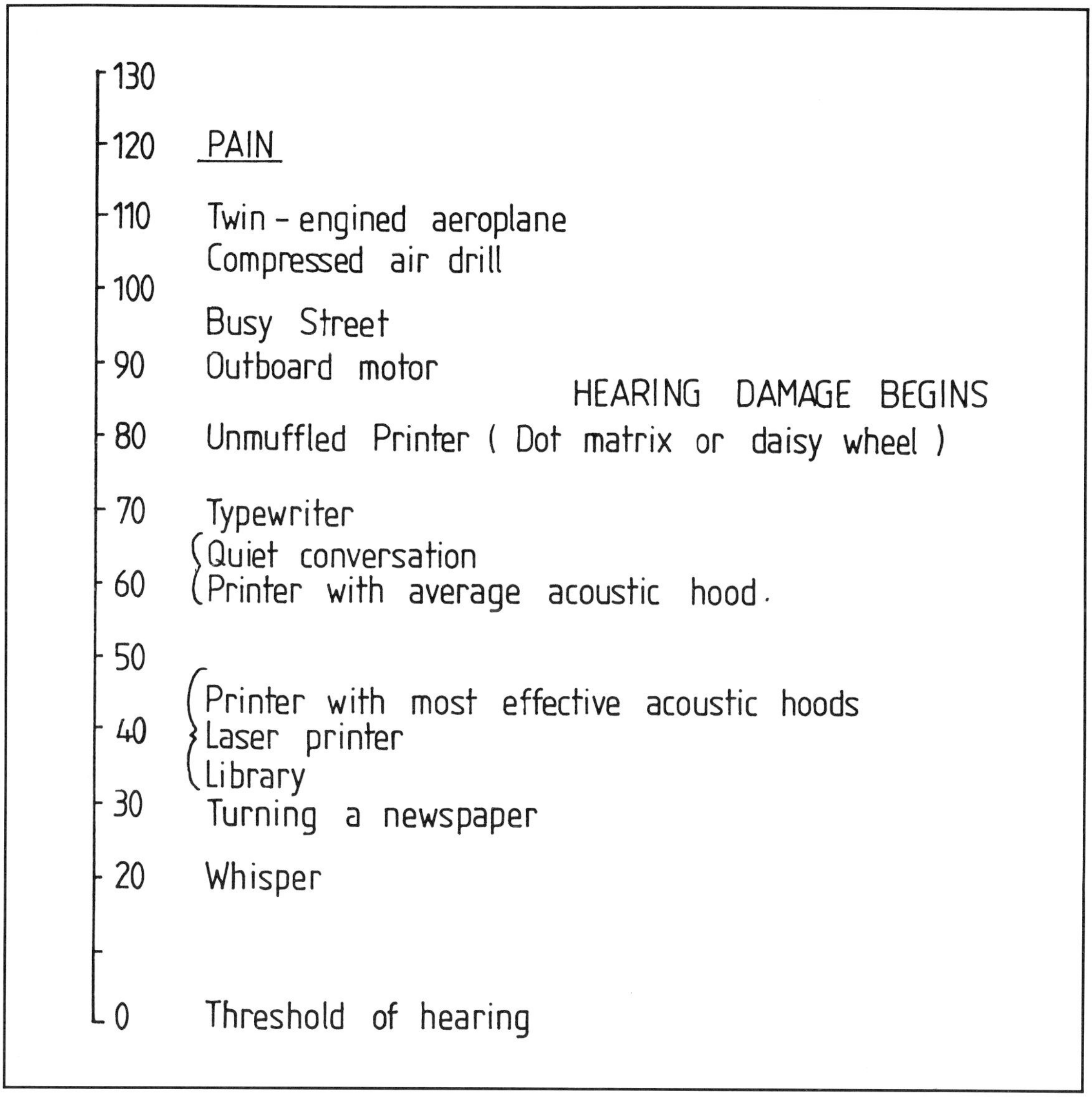

Figure 11.1 To show approximate equivalents to the dB(A) scale

- One's involvement with the noise – other people's conversations can arouse curiosity and therefore be distracting; the mention of one's own name calls attention more than the rest of the conversation.
- One's control over the noise – a sound which is imposed on a person is more annoying than one which can be controlled.
- The degree of acceptance of and adaptation to the noise – people vary in their ability to concentrate and to shut out noise and different kinds of noise affect people in different ways: a ticking clock, for

example, can be calming to some, ignored by others and annoying to a third group of people.

The rest of the chapter examines the problems caused by noise, the causes of noise, how noise can be reduced and, finally, the recommended noise levels.

The problems caused by noise

As with our other senses, hearing is for a purpose: to enable communication between us and other people and to act as an alerting device. Since the human brain responds particularly well to sudden or intermittent noises, such as the ringing of a telephone (see Chapter 1), such noises can be a source of distraction which can, in turn, lead to stress, especially if the work being performed requires continuous attention.

In order to communicate effectively, we need to be able to hear and to distinguish what other people are saying. In order to hear speech clearly, it must be 10 dB higher than the level of the background noise and, if the subject is new or contains unfamiliar words or if the speaker has a difficult accent, the voice needs to be 20 dB higher. This is likely to be especially true, too, of conversations with people who cannot be seen, such as telephone conversations. The apparently illogical but surprisingly common statement, 'I'm sorry I can't hear you properly, I haven't got my glasses on' may be due to the fact that, when hearing is difficult, the listener often unconsciously supplements the hearing loss by lip-reading.

Continued exposure to certain levels and types of noise – for example working with a compressed air drill – can cause hearing damage. Work in libraries and offices is unlikely to cause deafness but, if much noisy equipment is used in poorly designed acoustic environments, there may be some risk. For the most part, though, the problems of noise experienced in automated libraries will be interference with verbal communication, loss of concentration and stress.

Removal of all noise however is no panacea. Apex (1985) reports that, in open plan offices in Sweden, considerable attention was given to eliminating all sources of noise. The result was that the staff were found to be suffering from a state of shock and it was necessary to reintroduce an acceptable level of background noise. 'White noise' generators, which produce a low level of background noise, can be used for this purpose.

The causes of noise

Noise is cumulative, and may be made up of a number of contributing factors: people talking on the telephone; people talking across the office; constant movement of people; ringing telephones; typewriters in use; printers; air conditioning and heating systems; cooling fans in VDUs and acoustic hoods – especially if they become worn and/or faulty; high-frequency noises from the cathode ray tube; photocopiers; office fans (in summer); and unrelated noise such as traffic, not all of which relate directly to automation but which are all likely to be present in a library or workroom and will add to the overall noise level.

In addition, noise, like light, can be reflected or absorbed. Noise bounces back from floors, walls, ceilings, work surfaces and metal office equipment such as filing cabinets. Hard surfaces, such as metal and stone, cause the sound to be reflected much more than do absorbent surfaces, such as carpet and hessian wall coverings. The shape and layout of the room also affects its acoustic qualities. A long narrow room will seem louder than a square room of similar area, because the sound has only a short distance to travel before being reflected and, as noise decreases over distance, it hits the wall of the narrow room at a greater volume than it would in the square room. Low ceilings have a similar effect.

The layout of the room dictates whether sound will be encouraged to carry or to be blocked. In one library, the room was arranged so that the bookshelves fanned out from a central issue desk. The purpose of the arrangement was to ensure visibility throughout the library to increase security. Where light could travel, so also could sound, and noise from the busy counter ran unchecked to the 'quiet' study areas.

How to reduce the effects of noise

The effects of noise can be reduced in several ways or by a combination of approaches. These include reducing the noise at its source or on its way to the hearer, and minimizing an operator's exposure to the noise.

Reduce the noise at source

The best solution, if possible, is to reduce the noise where it is generated. This may be done in one of the following ways.

- Ensure that equipment is regularly checked and maintained so that cooling fans and so on do not become noisy.
- Obtain an acoustic hood for the equipment or use a pad under the printers to reduce the noise (see Figure 11.2). Misco (1988) claims that up to 90 per cent of printer noise can be reduced with an acoustic hood. Certainly the decibel level is lower (see Figure 11.1) but hoods which fully enclose the printer may need a fan to ensure that the printer does not overheat. The fan itself needs to be checked for noise.
- If buying a new printer, find out from the manufacturer the working noise level, but be aware that it is likely to be the average rather than the peak level and will not take into account the environment in which the printer will be situated.
- Isolate noisy equipment by, for example, putting the printer in a separate room. However, this may be inconvenient, if people need frequent access to the printer or a constant change of stationery is required – for example, to replace labels with catalogue cards or printed stationery. Problems may also arise if an operator is forced

to work in isolation, not so much because of the equipment noise, but because of loneliness.

Figure 11.2 An acoustic hood in use on a public access printer. Photo by Rick Dowling

Reduce the noise on the way to the hearer

This means, in effect, putting a sound-absorbent barrier between the source of the noise and the operator(s) who will be affected. Double-glazing can cut down external noise such as traffic but, in summer especially, it will be important to ensure effective cooling systems as opening the windows for air could also admit noise.

In addition, the environment can be made to absorb sound generated from inside or outside the building. Any of the following sound-absorbing materials could be used: carpets on the floors; acoustic ceiling tiles; partitions (shelves of books are extremely effective but care must be taken to ensure that operators do not feel isolated); absorbent paint or materials to coat the walls.

Reduce the period of exposure to the noise

While some work may have to be done in a noisy area, it may be possible to allocate times away from distractions to undertake work requiring greater concentration. Lack of noise (or other) interruptions may enable the operator to work more effectively, thus attaining a greater sense of job satisfaction.

Recommended noise levels

Apex (1985) suggests an office noise level of 65 dB(A) but argues that a lower level should be aimed for. The ANSI standard cited by Czaja *et al.* (1987) recommends a level of 55 dB(A) and the BP Group Occupational Health Centre (1987) claims that the ideal is 45–55 dB(A). In an open plan office or library, 50–55 dB(A) is to be aimed for. Ideally, this should be made up of steady, unchanging sounds such as the air conditioning system and the hum of a computer. Such sounds help to mask other people's conversations, which can be distracting. Less than 45 dB(A) allows other sounds, such as speech and the rustle of paper, to travel further, and this can be annoying.

An investigation undertaken by Czaja *et al.* (1987), however, indicates that noise level need not be the biggest problem. Czaja and her colleagues studied 105 female workers in both traditional and automated offices. The average noise level was found to be 51 dB(A); the range was not stated. Of those studied, only 31 per cent found the office to be too noisy, yet 66 per cent were annoyed by talking, 42 per cent by telephones ringing and 28 per cent by printers and keyboards. So, although only about one-third of the workers felt that the overall noise level was too high, two-thirds were irritated by a single type of noise – talking.

Thus, rather than simply reducing the overall level of noise, it may be advisable to obtain employees' views on the various acoustic distractions which they may experience. This is best done over a period of time to allow for the daily, and even hourly, fluctuations.

12 HEATING AND VENTILATION

The climate of the library is affected by four principal factors:

- air temperature;
- radiant heat;
- relative humidity;
- air movement.

These all affect each other and cannot be considered in isolation; nevertheless, for ease of presentation, they will be discussed under these four headings.

Air temperature

Individuals vary in their preference for heat and cold, and it is impossible to please everybody. However, recommendations have been made that the air temperature should be maintained between 21° and 23° centigrade (C) for light office work and a legal minimum of 16°C has been set, below which it cannot be expected that work will be undertaken (Damodaran *et al.*, 1980).

In most modern library buildings, thermostatically controlled air conditioning is installed, bringing hot or cold air into the library as required to maintain a constant temperature. Both VDUs and people generate heat (approximately 100–300 watts per unit) and it is possible that, if VDUs and their operators are concentrated in one section of the library, as OPAC terminals often are, a 'hot spot' will occur in that area. It is therefore advisable to ensure that VDUs are well spaced throughout the workroom or library and to choose VDUs with low heat emission. When positioning the VDUs throughout the library, it is also recommended that heat output from the vents should not be directed towards the operator or his neighbour.

The different kinds of work undertaken in a library create their own problems for the heating system. Active work, such as shelving, requires moderate physical exertion, causing the person to generate metabolic heat. The shelver will therefore require a lower air temperature than the sedentary worker. Variations in the routine of the VDU operator need to be made for postural reasons (see 'How to minimize postural problems', p.31). It may also be necessary to consider the effects of long-term sedentary work on the operator's thermal comfort.

Radiant heat

In non-air conditioned libraries, heating may pose an even greater problem than in air conditioned buildings. Cakir *et al.* (1980) found that 50 per cent of people in traditionally heated rooms complained of problems relating to heating, compared with 30 per cent of those in air conditioned environments. Certainly, it is in the non-air conditioned libraries that many of the problems of radiant heat occur.

The term 'radiant heat' is used to refer to heat emitted from a surface or a body, such as a central heating radiator, a light bulb or a VDU. Thermal discomfort may occur if the heat (or cold) emitted from the item is considerably different from the air temperature (±50 per cent). It may be necessary to rearrange the furniture and equipment to ensure that the operator's legs, for example, are not situated uncomfortably close to a radiator.

Figure 12.1 Potential for thermal discomfort. A VDU workstation adjacent to a radiator and in full sunshine.

Another form of radiant heat is the sun, especially through south-facing windows. Effective curtains or blinds must be available to cut out the unwanted sunlight when required. A combination of a radiator and sunshine could be particularly uncomfortable (see Figure 12.1).

Relative humidity

Relative humidity is a measure of the amount of moisture in the air and is expressed as a percentage. Cakir *et al.* (1980) recommend that the relative humidity should not fall below 50 per cent and should not fluctuate widely. It is suggested by some that the comfortable range is between 30 and 70 per cent relative humidity. Nevertheless, two-thirds of operators in a survey reported by Cakir and his colleagues commented that a level between 30–40 per cent was too dry. Thus a level higher than 40 per cent is recommended.

If the atmosphere is too dry and, especially, if it is also too hot, staff may experience dry noses and throats. The influence of a dry atmosphere in causing facial rashes has already been discussed (page 67). The relative humidity can be increased by reducing the temperature or by improving the air conditioning system so that it also humidifies. A simpler and cheaper solution is to use humidifiers which clip into the radiator or to distribute plants around the library. Electric humidifiers can also be purchased.

Some air conditioning systems which also humidify have been suggested as the cause of a new occupational disease, 'Sick Building Syndrome'. The symptoms include intermittent chills and fever, coughing, vomiting, aching joints and tightness in the chest. It has been suggested, however, that the cause of this problem is low morale rather than medical (Spackman, 1988). More commonly, too high a relative humidity can cause stuffiness. Dehumidifiers may be purchased, or the air temperature can be increased to reduce the humidity level.

Air movement

Ideally, air movement should be at 15 cm per second; below this speed, the environment is perceived as airless and above it, as draughty. In one study by Cakir *et al.* (1980), draughts were found to cause approximately 25 per cent of neck ache complaints. They were also found to be more prevalent in air conditioned environments, especially during hot weather when air currents are used to cool the atmosphere. Air movement is more readily perceived if the air is cool, so if the air temperature is raised, its movement is less noticeable.

In hot weather, however, fast air movement may be desirable. A rapid flow of air over the body encourages heat loss by evaporating perspiration; hence the use of fans in summer. Opening windows is another way of encouraging increased air movement but brings with it the hazards of admitting noise and fumes as well as causing disturbance to piles of documents.

Recommendations

A comfortable climate is necessary to create a good working environment, and the following actions are suggested.

- Elicit from the staff their feelings about the climatic conditions.
- Look, in particular, for 'hot spots' and sources of radiant heat and reorganize the area if necessary.
- Where possible, obtain VDUs and other equipment with low heat output, and position the equipment so that the operator, neighbouring operator or library user are not adversely affected by the heat output vent.
- Ensure effective curtains or blinds are available to shut out unwanted sunlight.
- Regulate the relative humidity by the use of humidifiers or dehumidifiers as necessary.

Part V WORKPLACE DESIGN: WORKSTATION LAYOUT AND DESIGN

13 WORKSTATION LAYOUT AND DESIGN

Introduction

A workstation is the combination of furniture and equipment used by an operator to carry out his job. The aim of good workstation design is to select and arrange the appropriate components in order to facilitate efficient operation and employee comfort, well-being and safety. Consideration has already been given to how to adjust a workstation to attain a good posture and minimize postural discomfort. This section considers how best to organize the various components with a view to efficient operation of tasks.

In particular, discussion centres on analysing a task to assess its design requirements, equipment considerations, the physical limitations of the operator, the principles of workstation design and its effect on productivity.

Task analysis

Before a workstation can be designed, it is necessary to understand how it will be used. This implies the need to study the various functions of the workstation in some detail, a stage known as task analysis. The objective of the analysis is to ascertain how frequently and for how long a particular task is to be performed. The more often a task or sub-task is performed or the longer it takes, the more consideration needs to be given to exactly how it is performed and to design the workstation to maximize comfort and efficiency in its operation.

The nature of libraries is such that no rules can be laid down as to how a particular workstation should be organized. Libraries vary both in size and the way in which tasks and sub-tasks are allocated. In larger libraries, for example, it is possible that a workstation can be specialized for a particular task: cataloguing, for example, may always be done by one individual at one terminal. In smaller libraries, however, many different functions will need to be performed by the same person at the same desk, and so a compromise may need to be reached. A workstation to perform the same function may also have different requirements in different libraries because of varying library policies. Rea (1987, p. 20) gives the following example of the effect of the University of Bath's library policy on the issue desk:

> In our library there is no restriction in the number of books a reader may borrow. Because there is no restriction, there is no compulsion for readers to wait at the desk while their books are discharged before they can borrow further books. Space, therefore, needs to be set aside where books can be piled up; hence the larger the normal size of our returns counter.

Task analysis considers how the workstation is used and the various tasks which are to be performed. The operator's posture, his interaction with other people and the amount of movement needed are also examined. In addition, the use of ancillary equipment, notices or operating instructions, and the changes brought about as a result of automation are studied. Each of these is discussed below.

How the workstation is used

While it is possible to analyse how the workstation is currently used, it should be noted that automation may bring about a change in the task. This may be because it will need to be performed in a different way with the automated system, or the job may need to be redesigned (see Part VII) to optimize operator performance. The library manager, in analysing the task, needs to consider a number of factors:

- The number of users. In the libraries studied by the authors the number of users using a single workstation varied from one to six.
- The age, sex and fitness of the users. The anthropometric data in Table 1.1 indicate the range of sizes to be catered for. The desk height (see pages 85–89) needs to take into account the tallest user (or 95th percentile) and reach requirements (see 'Reach' p.139) the shortest (or 5th percentile).
- Whether the users are right- or left-handed.
- The length of time spent by each operator and on each task. The time spent at the VDUs may vary from a few minutes at infrequent intervals in one library to several hours at a stretch, several days a week in another library.
- The frequency with which each task is performed.
- The frequency with which the workplace is used.
- The number of different tasks performed.

The tasks performed

The tasks to be performed at a workstation will differ according to individual circumstances. It is therefore essential to examine them in detail, bearing in mind such factors as the proportion of VDU and non-VDU work, the category of VDU work and the sequence of activities.

The proportion of VDU and non-VDU work

In small libraries, for example, the operator may fulfil all the functions of the library, creating catalogue entries, ordering books, answering reader queries, doing online searching and so on. Such a range of tasks will inevitably include a mix of VDU-based and non-VDU work. The proportion of time spent on each will influence the amount of space required for each type of task. The use of documents or manuals in conjunction with the VDU will require adequate space, especially if two or more need to be read in conjunction with one another as may be the case in online searching. Space may be needed to sort documents, such as overdue notices prior to posting, to make piles, such as books for cataloguing or serials for check-in, or to make notes.

The category of VDU work

This may be loosely divided into the following: data entry; data enquiry; interactive or conversational; and word processing. The distinction is drawn between each according to the relative use made of the three main components in VDU work, viz. the screen, the keyboard and the source document.

Data entry is the term used to describe continuous input of data, such as retrospective catalogue conversion from the catalogue cards or the input of a document to the word processor. The operator, for the most part concentrates his gaze on the source document with an occasional glance at the screen and keyboard. Studies have indicated that the percentage of time spent looking at each of the three main components is: 70–80 per cent on the source document; 15–20 per cent on the display screen and 3 per cent on the keyboard (Grandjean, 1987). Keying is usually rapid and operators tend to adopt a fixed posture. Ideally, the items which are to be viewed most frequently need to be situated near to, and at the same angle as, one another. In the case of data entry, the source document needs to be in the prime position with the screen nearby and at a similar angle (see Figure 13.1a).

It has been found that, for data entry, operators using a straight-line VDU arrangement, similar to that illustrated in Figure 13.1b were more efficient and more comfortable than with a lateral arrangement as in Figure 13.1c (Marek *et al.* 1984). Grandjean (1987), however, argues in favour of positioning the source document in front of the operator with the screen to the right-hand side and the input pile to the left (for right-handed operators) as shown in Figure 13.1a. Operators with personal preferences should have the last word.

Data enquiry involves calling up information from a screen and reading it. This is the kind of operation which is performed on an OPAC terminal or on a Viewdata system. Little use is made of the keyboard, the operator spending most of his time reading from the screen, possible searching for specific items of information such as exchange rates or flight times on Prestel, rather than reading the entire screen. It may be necessary to note down what appears on the screen.

An interactive, or conversational, task combines a mixture of data input and data enquiry, further input often being required in response

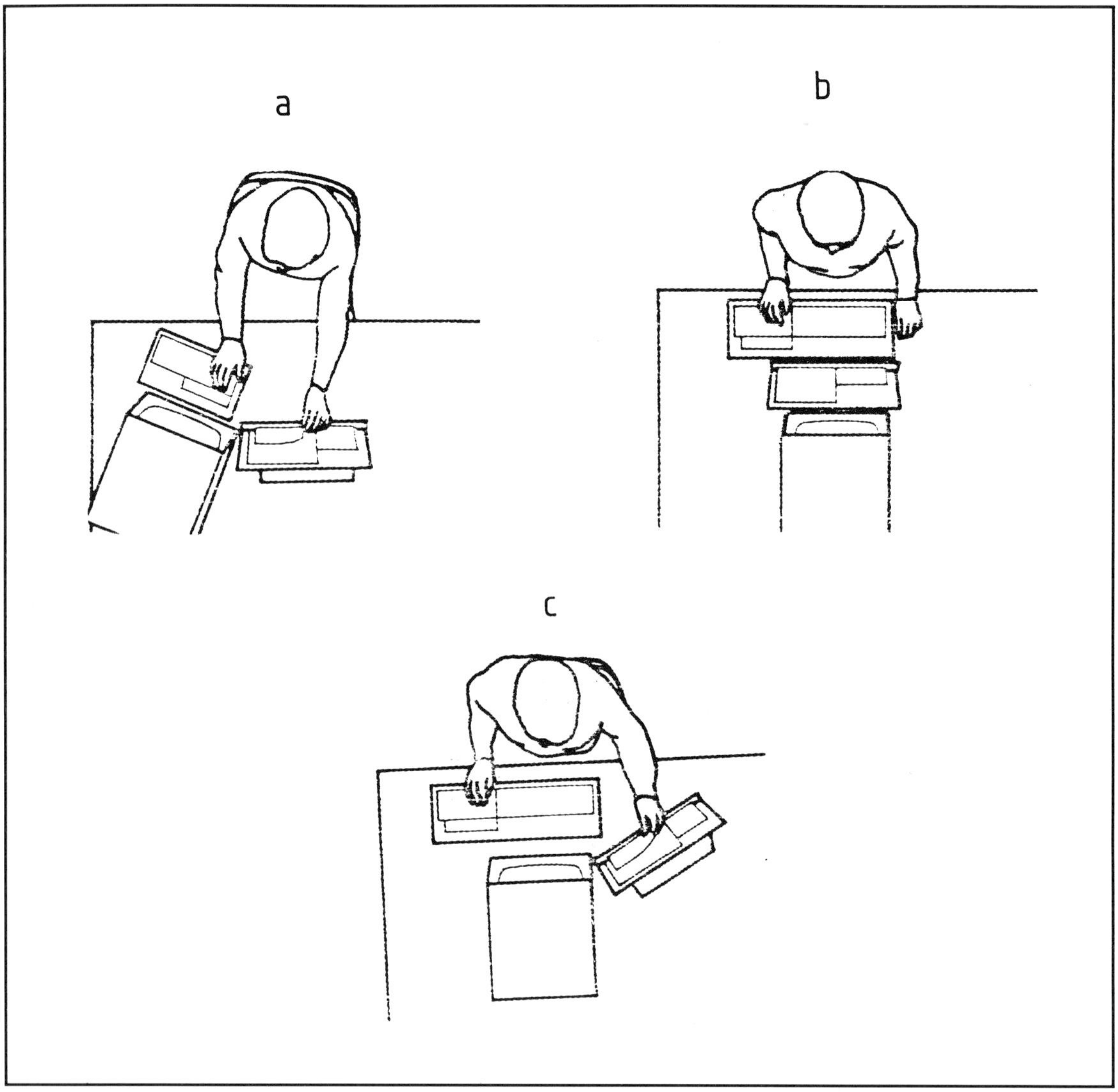

Figure 13.1 *To show three suggested layouts of workstations.* Adapted from Health & Safety Executive, 1983

to the screen display. A good example is online searching. As it is likely that the operator is not a touch typist, keying speed will be low and the operator may have to look at the keyboard while typing. Grandjean (1987) reports the following percentages of time spent looking at the three elements: 35 per cent at the source document; 30–40 per cent at the display; and 20–35 per cent at the keyboard. Thus the three elements should be positioned so that they are equally easy to see. Grandjean recommends placing them in a straight line, as in Figure 13.1b but if, as with online searching, it is necessary to write notes, it is preferable for the form to be positioned on the right (or left) rather than behind the keyboard.

Word processing, or to be more precise, the amendment of text on a word processor, demands a considerable amount of time looking at the screen, with less time each at the keyboard and the source document.

The sequence of activities If there is a logical flow to the work, items need to be arranged to facilitate ease of movement between them.

Posture Posture is discussed in depth in Chapter 8, pages 77–80. For the purposes of task analysis, it is essential to ascertain whether the operator(s), for whom the workstation is to be designed, will be sitting, standing, or a mixture of both. (The last is to be encouraged where practical.)

Interaction with other people A variety of forms of interaction may occur. There may be interruptions from colleagues, readers may be present at an online search and work at the issue desk will usually involve interacting with readers. The direction from which these interruptions come is important. If interaction is to be encouraged, it is desirable for the operator to be facing towards the likely source of the interruption. Becket and Smith (1986) describe their dilemma in designing a reference station from which staff could view a 360° angle, in order to encourage users to approach from any direction.

Interaction with users for online search purposes requires a position which facilitates communication between the user and the searcher but which excludes unnecessary interruption. Alternatively, it may be necessary to minimize unproductive interaction between staff, but without inducing a sense of isolation.

Movement If frequent access or egress is required, it is necessary that equipment is designed and space is available for ease of movement; chairs on castors, for example, are easier to get on and off than chairs on legs, and armrests can be obstructive. One of the features of library work, particularly at the counter, is the mobility of staff.

Ancillary equipment Frequent telephone calls dictate easy access to the telephone. Similarly, other ancillary equipment such as wastepaper baskets, book trolleys, cash registers (for fines on the issue desk) need to be easily accessible. The frequency of use of the ancillary items can be used to rank them in order of importance if there is competition for space.

Notices for operating instructions Instructions for use of an OPAC or for end-user online searching need to be given prominent space and need to be eyecatching. Less frequently used or more confidential instructions, such as identification numbers and passwords for online searching, need to be secure but easily accessible by authorized users.

Figure 13.2 The reference desk at the University of Rochester Library, Rochester, New York, designed to encourage users to approach from any direction.
Source: Becket and Smith, 1986. Photo by Jeffrey Goldberg, University of Rochester, reprinted from *Library Journal*, April 5, 1986. Copyright © 1986 by Reed Publishing, USA, Div. of Reed Holdings, Inc.

Changes caused by automation

A change from a manual to an automated system may itself impose a change of sub-tasks. Barcoded entry at the issue desk, for example, reduces the time spent issuing and discharging books. Rather than allowing periods of enforced idleness for staff, other work may be introduced as an alternative task, bringing with it its own requirements for workstation layout.

A change of desk layout may also have an effect. Becket and Smith (1986) describe how the transfer of the reference desk to a more prominent position in the library brought an increase in reference queries.

Equipment considerations

As well as understanding the nature of the work to be performed, it is necessary to consider the computer equipment which will be used. Margaret Beckman (1987) urges consideration of the following questions:

- What are the electrical, mechanical and communications features required?
- What is the likelihood of this technology changing so that the equipment will be replaced by something new, either larger or smaller, and with different requirements? (Somerville *et al.* (1986) decided against sinking terminals into recessed wells when designing an issue counter in order to avoid obsolescence were future equipment to be of different dimensions.)
- Will this piece of equipment require staff assistance in its use – or surveillance for security purposes?

Operator limitations

It is necessary to place objects and materials so that the operator can reach and see them comfortably. An operator's physical limitations need to be appreciated when designing the workstation.

Reach

If an employee is forced continually to reach too far to operate a control or to pick something up, this can lead to pains in the back and shoulders, especially if the object being retrieved is heavy, like a book. As Table 1.1 shows, people vary in stature and, consequently, in their reach. Reach is primarily in two directions – vertical and horizontal – and these are discussed below.

Vertical reach

The limit of upward or downward reach is the radius of the arm, up or down, with the hand in a clasped position – that is, the distance from the shoulder to the fist. Unlike the desk heights, which cater for the tallest employee, reach requirements should be based on the smallest employee. Table 13.1 gives the relevant measurements for the 5th percentile (the measurements of the smallest 5 per cent of the population).

The highest position that can be reached comfortably from a sitting position is 107 cm for small women and 119 cm for small men. Allowance may need to be made for children, if appropriate. This applies only to vertical reach, however, and does not take into account any forward movement. If a shelf is in front of the operator and the required item at the back of the shelf, the distance from the operator's shoulder to the item should not exceed 107 cm, which would mean that the shelf should be lower to compensate for the forward distance.

Table 13.1 The 5th percentile measurements relating to reach. Adapted from Grandjean, 1980

	Small Men (cm)	Small Women (cm)
standing shoulder height:	130	120
sitting shoulder height:	54	49
arm length (closed hand):	65	58
max. reach from sitting:	119	107
max. reach from standing:	195	178

Horizontal reach

The various reach distances are as follows (Grandjean, 1980):

35–45 cm – relaxed arm – optimum reach
55–65 cm – straight arm – maximum reach
70–80 cm – stretching, using the whole body

They are shown diagrammatically in Figure 13.3 below.

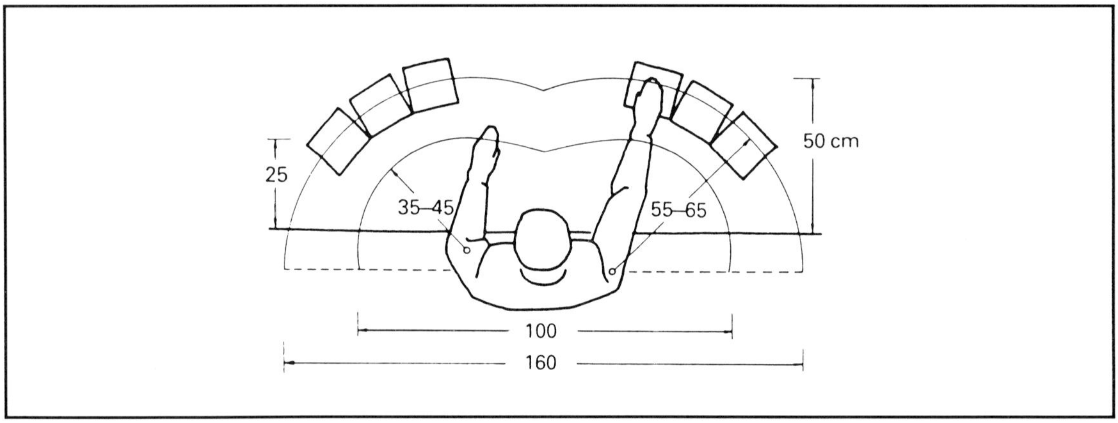

Figure 13.3 The maximal and optimal lengths of horizontal reach
Source: Grandjean, 1980

Forward, horizontal reach may be divided into two categories – optimum reach and maximum reach (Shackel, 1983). Optimum reach is defined as being the area which is easy to reach with the elbow resting on the desk – that is, the arc from the elbow to the hand. Writing, for example, is usually performed within the optimum reach 'envelope'. Maximum reach describes the arc of the outstretched arm – that is, from the shoulder to the hand, in much the same way as for vertical reach. Beyond the maximum reach envelope, it is necessary to stretch the body.

The least tiring arm position is within the optimum reach envelope where the elbow can give support. Maximum reach is unsupported, the shoulder and arm muscles carrying the load, not only of the outstretched arm but also of any items which may be being carried. Thus a weight in the hands diminishes reach (Pheasant, 1986).

Thus, ideally, the most frequently used items should be within the optimum reach envelope. The following equipment, among others, should be considered: the VDU screen, so that its angle can be adjusted if necessary; the controls – ON/OFF, brightness, contrast, modem; source documents, such as books; paper for writing notes etc.; telephone; pens/pencils etc.; floppy disk drive; waste bins; and personal belongings.

Figure 13.4 shows the effect of the confines of one library workroom. Note that all the items which may be required need to be reached by twisting round or by stretching upwards.

Figure 13.4 Lack of space results in items being stored out of comfortable reach. Here, the operator must stretch up or twist round to reach manuals. Photo by Rick Dowling

The position of the equipment can be laid out according to four methods (Damodaran *et al.*, 1980, p. 86):

- by importance – where the controls and items of equipment considered the most important occupy the prime positions;
- by function – where equipment relating to one function is grouped together separately from that relating to other functions;
- by frequency of use – those items used most frequently are located in the optimum reach envelope, less frequently used items are situated in the maximum reach area;
- according to sequence of use – such as stuffing envelopes for a mailshot or collating papers – in order to reduce the risk of error. A regular order, preferably from left to right and from back to front or top to bottom, is advised.

View

The distance that an operator can see will depend on several factors, such as the quality of his eyesight, the size and clarity of the characters on the VDU screen or source document, the quality of the lighting and the effect of any glare. The recommended distance between screen and eye is 61–93 cm (Grandjean, 1987).

The angle of view will depend on eyeball movement and the tilt of the head and neck. The eye can move within 15° above and below the normal line of sight. In order to see something at a greater angle, it is necessary to move the head and neck.

The normal, or preferred, line of sight is 10–15° below the horizontal plane. The arc of view made possible by eye movement gives a comfortable range of sight from 5° above the horizontal to 30° below it, without moving the head or neck. Thus, to reduce the risk of painful neck and shoulder muscles (outlined in Chapter 3), all items which are to be viewed frequently should be positioned so that they may be seen with minimal head movements. Ideally, therefore, items should be at the same angle and the same distance from the operator and situated close together where possible.

The angle of the screen and the source document should be the same and recommendations are that the screen angle should be between 75° and 110° to the horizontal (Grandjean, 1987). The precise angle of the screen will depend on the operator's position relative to it. Ideally, the screen should be at right angles to the operator's normal line of sight.

Incorrect positioning of the screen can cause the head and neck to be held at an awkward angle. The angle between the neck and head and the rest of the body should be between 0°and 15° and should not exceed 20–30° for a prolonged period. Looking up for too long, causing a backward tilt of the head beyond 30°, is extremely uncomfortable, and discomfort has been found to be greater when the screen is too high, rather than too low (Grandjean, 1987). Rea (1987, p.18) describes the effect of this problem on the staff of Bath University when automated equipment was placed on an issue desk designed for a manual issue system. The VDUs were

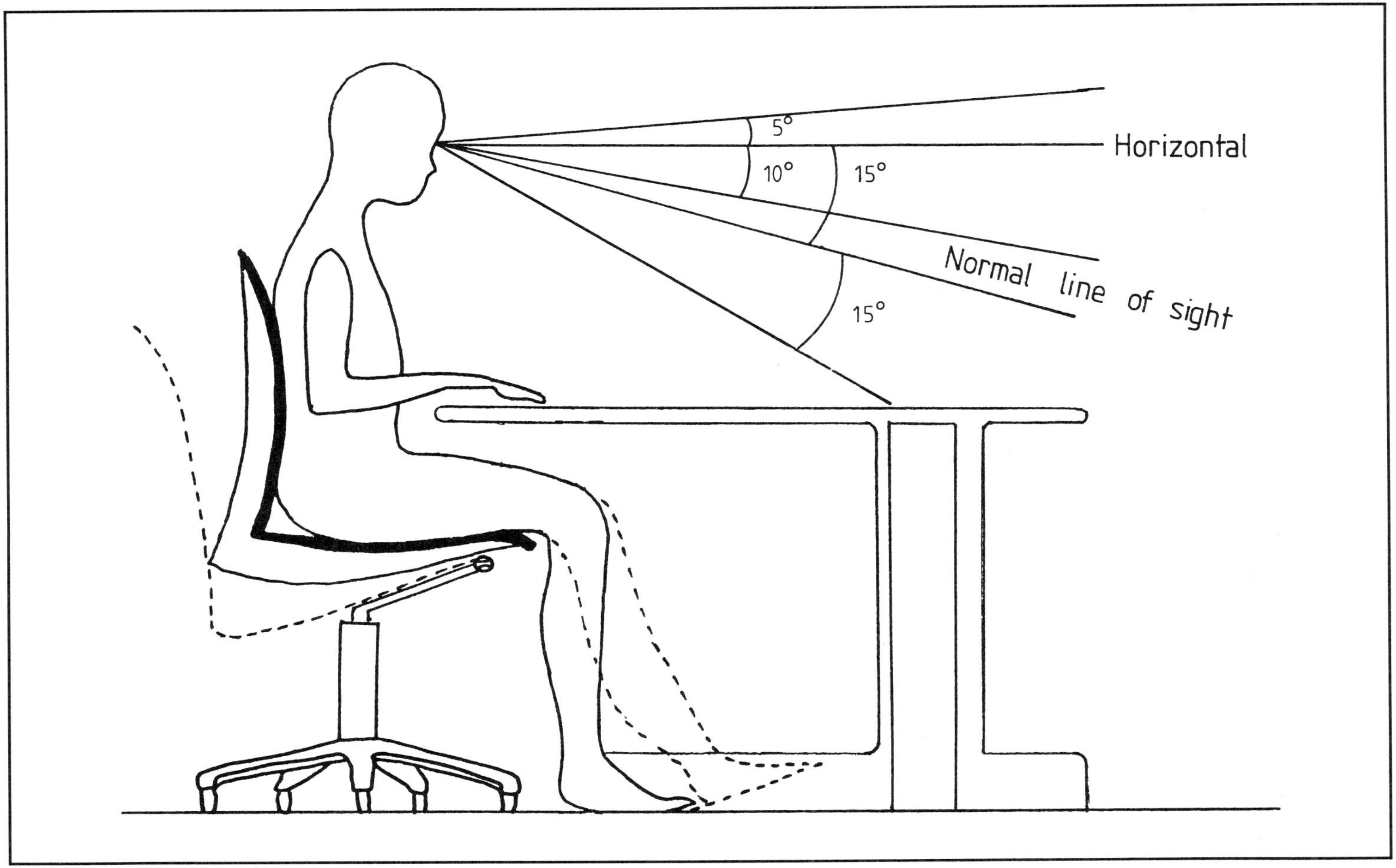

Figure 13.5 Operator viewing angles

perched on the ledge which was at standing height for readers and the staff, who were sitting below the level of the VDU, had to look up to see the screen. This 'caused stiff necks and eye-strain (the alternative of standing and leaning over was just as bad)'. More effort is expended when the head is moved in the vertical plane, such as when nodding than when it is moved horizontally (Cakir *et al.*, 1980). Ideally, therefore, the source document should be held on the same level as the screen to permit side to side, rather than up and down movement.

In order to position the source document at the same angle and distance as the screen, a copy holder may be used. This is particularly useful for data input such as word processing or cataloguing from a source document. It should be noted, however, that the more the angle between the lighting and the document is reduced, the less clear that document may appear. However, Damodaran *et al.* (1980) recommend the use of a document holder to reduce the luminance contrast between the screen and the source document.

A range of copy holders is now available:

- Those that can be attached to either side of the monitor, offering a narrow viewing distance between the screen and the source document and a similar angle. These can be purchased for anything from £7 to £35.
- Freestanding models made of perspex or acrylic. These may be able to accommodate A4 or even A3 paper, but are not designed for books. This kind usually costs less than £20.
- Clamp-supported or heavy-based holders with an angled arm bracket. Once again, these are not appropriate for books, but will hold A4 or A3 paper. They cost anything between £30 and £80.
- Electrically operated copyholders with foot pedal adjustment, which cost from £90 to over £160.

With such a range, it is useful to know what features to look for. The following criteria may be used:

- There should be a low reflectance surface to minimize reflected glare.
- It is useful to be able to alter the rotation, distance and slope of the holder.
- A reading guide or cursor line is helpful.
- It should be possible to adjust for varying thicknesses of source documents.
- It should be adjustable by the operator to the right or left of the screen.

Most copy holders are designed for single sheets or small pamphlets and few offer the necessary support or allow for the thickness of books. This may be overcome by the use of a lectern but, in one library, it became necessary to have a copy holder specially made. This was to assist in the cataloguing of heavy reports, many of which had long titles or large numbers of authors, and so required the assistance of a cursor line.

The type of work performed will also affect the type and position of copy holder which is best suited to the task (see Table 13.2).

Table 13.2 Manuscript holder recommendations for VDU tasks

Task	Frequency of Manuscript Change	Type and Position of Manuscript Holder for Right-hand Holders
Pure copy entry, no manipulation	Seldom	Size: according to type of document. Row marker: yes. Position: to the left of the display screen, ° angle
Copy entry with some manipulation, e.g. occasional notes	Seldom Row marker: yes.	Size: according to type of document. Position: to the right of the display screen.
Pure copy entry, no manipulation	Frequent	Size: according to type of document. Row marker: preferably. Position: to the left of the display screen, or between the keyboard and the screen
Copy entry with some manipulation	Frequent	Size: according to type of document. Row marker: preferably. Position: to the right of the display screen.
Pure copy entry, no manipulation	Very Frequent	Size: according to type of document. Row marker: no
(Mostly numerical data)		Position: to the left of the keyboard

Source: Damodaran *et al.*, 1980

Three main items therefore need to be comfortably in view: the screen, the keyboard and the source document (or paper for notes). The relative position of these to one another needs to be such that the correct angle of view is maintained and unnecessary head and neck movements are minimised. As with reach, view must take into account the relative frequency with which each element is observed.

Workstation design

In order to permit efficient operation, workstation design must take into account factors such as the task to be performed and the physical limitations of the operators. Other factors which must also be considered are: the need to adjust the system should requirements change; storage requirements; ease of cleaning; the amount of worksurface required; the need for cabling; maintenance, and security; the preferred construction material; and the effects of static electricity. These are discussed below.

Flexibility Beckman (1987, p.16–2) argues: 'There must be flexibility in the design of furnishings which house the equipment because the equipment size and communications connections may change.' Flexibility for the different sizes and shapes of employees is also important. However, they may need to be educated in how to adjust a workstation to suit individual requirements, as experience indicates that staff do not usually make adjustments to furniture (Rea, 1987; Ralli, 1988). Flexible, modular and adjustable furniture is now easily available (see the list of suppliers in the Appendix). The layout must allow for operator preference in terms of posture, arrangement of items, and performance of the task.

Storage requirements There should be enough storage space to hold all the items needed to perform the task, including manuals, spare paper, forms, trolleys or shelves for books awaiting or following processing, and personal items, such as handbags. The need to be able to access storage easily should be taken into account, as should the need for locks. In many libraries, lack of storage space is a problem.

Ease of cleaning A dirty library or workroom is bad for public relations and staff morale. Dirt traps, such as poorly designed drawer handles should be avoided. The use of dust covers for VDUs and keyboards is also recommended for when the equipment is not in use. They can also protect the equipment from other hazards, such as coffee spills.

Worksurface area If a VDU is placed on an existing desk the usable space on that desk is reduced. Space needs to be available in front of the keyboard on which the operator can rest his hands and forearms. It may be necessary, too, to be able to push the VDU backwards or forwards to cater for different preferred viewing distances or to give more space when the VDU is not in use. Slovak and Trevers (1988, p.100) suggest that 'an area of worksurface at least 100% greater than with typical typing work needs to be dedicated to the job when VDTs are used, unless all transactions are paperless'. They found that lack of space was one of the biggest problems to the users they surveyed, complaints being recorded from 36 per cent of respondents.

The use of a 'buffer' area may also need to be considered for incoming and outgoing documents or books. For example, when checking in journals, there needs to be space for the waiting pile, the item being processed and the output pile. Similarly, when cataloguing, a small pile of books awaiting processing should be within easy reach of the operator to avoid having to twist and reach for a new item from the trolley or shelves each time. The need for a buffer area within comfortable reach may be met effectively by curving the worksurface around the operator. This may be done using specially designed or modular furniture (Figures 13.6 and 13.7) or by positioning trolleys close to the operator. The latter arrangement offers little space for access or egress, however, and so is unsuitable for tasks requiring constant movement to and from the workstations.

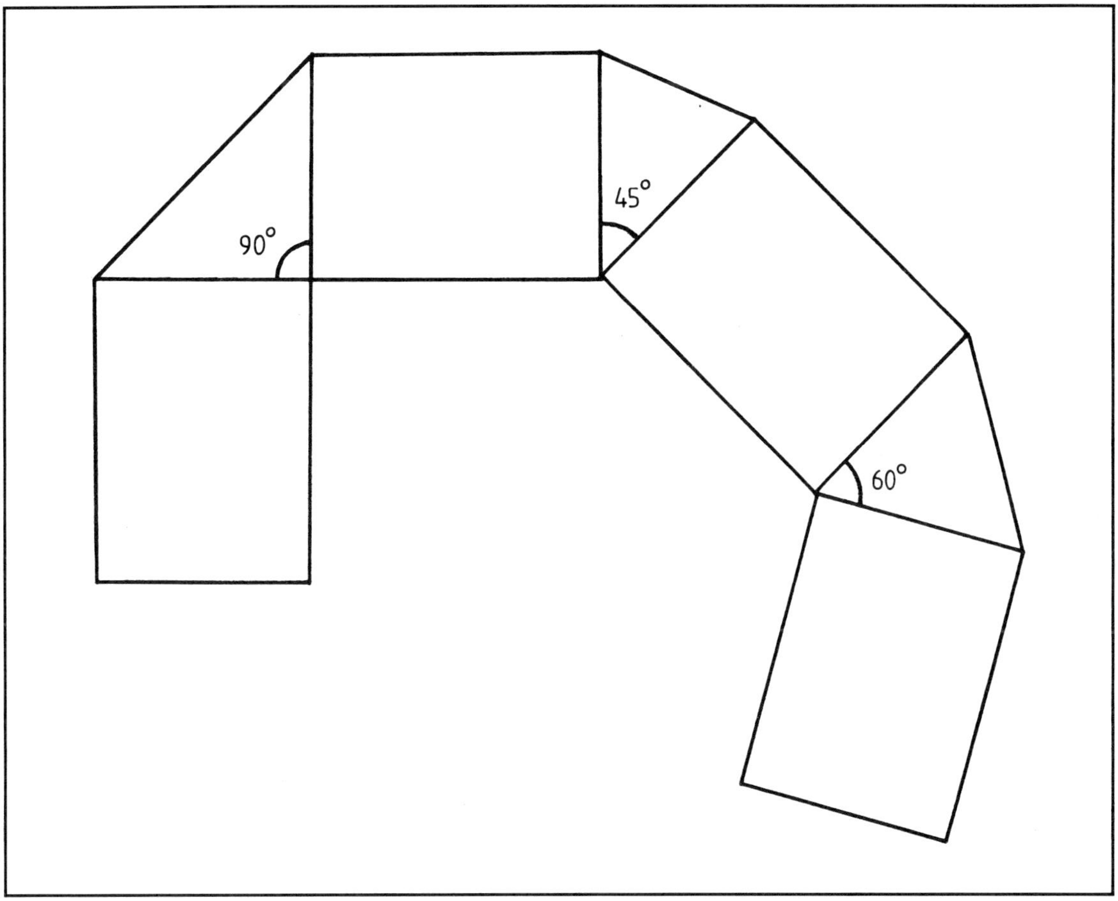

Figure 13.6 The use of modular furniture can offer an increased worksurface and a range of angles between adjacent surfaces

In order to increase the available working area, various space-saving devices can be used, such as:

- Folding arms which support the VDU screen. These can be adjusted at an angle to suit the operator but tend to position the screen too high relative to the operator. They do, however, make much more desk space available and even enable a terminal to be shared between adjacent operators.
- Keyboard holders which enable the keyboard to be slid out of the way when not in use. Again, the result may be that the screen is too high if the keyboard holder and CPU unit are situated underneath.
- CPU holders which house the unit on its side on the floor to one side of the desk. However, this is not without problems. The disk

Figure 13.7 *The use of modular furniture to extend the working surface within reach of the operator.*
Photo courtesy of Project Office Furniture plc

unit is below the comfortable level of reach of the operator, so it is not recommended for tasks where frequent disk changes are important. Also, the CPU and disk may be vulnerable to knocks and dirt, unless it is well positioned and covered. In addition, to position the disk drives on their side may affect the disk performance or could even cause damage to the drive.

Cabling

Loose cables under desks or on the surface of desks can be dangerous and since they could either get pulled or cause people to trip. This can lead to personal injury, damage to the connections (the pins are surprisingly easily bent), or even pull the equipment off the surface and onto the floor. Cabling is an important issue – indeed it was considered so important in Hillingdon's new library that trunking was said to be the 'backbone' to their IT effort (Westlake, 1988). Not all libraries, however, are in the position of being able to redesign the cabling from scratch. In addition, it would appear that few librarians are in a position to know how to plan for electrical and communication requirements (Somerville *et al.*, 1986), and it is suggested that the importance of providing adequate, safe cabling warrants the advice of an independent expert.

A few general points are worth noting here, however:

- Flexibility should be built into the cabling system to allow for adding or moving micros or terminals. Beckman (1987, p.16–3) argues that: 'Panels built into the workstation which hide the conduit or cables at either the base or work surface level can allow for future change. Equipment bars which group a variety of electrical and telecommunications features provide flexibility for groupings of equipment in confined areas . . .'. Rea (1987) used a slightly different method to hide the cables in the new issue desk at Bath University. A narrow channel was built into the front of each unit into which boards were slotted to protect the user from the back of the VDU (and vice versa!). These channels were also used to house cabling.
- Wires should be unobtrusive. A number of system units now offer discreet cabling inside desk legs, in channels in the desk and through screens or dividers (see Figures 13.8 and 13.9). A cheaper, but less attractive, alternative is to drill holes in the existing workstation to feed cables through, although future changes may require more holes to be drilled which may be unsightly.
- Accurate anticipation of future cabling requirements is important, although far from easy. Westlake (1988, p. 8) honestly admits that at Hillingdon mistakes were made causing the 'trunking [to be] opened four times, with the inevitable risks for previously soldered connections'. In particular, it is difficult to envisage the future requirements for public access terminals. Fayen (1983, pp. 41–2) urges libraries to think big when it comes to providing OPAC terminals:

Figure 13.8 *Cabling of all sizes can be accommodated through channels in some office systems furniture.*
Photos courtesy of President Office Furniture

Studies have shown that users want terminals – lots of terminals – and they want them not only in the library but in other locations. For a public library this might mean placing terminals near the entrance, in the children's room and in the stacks; it might also mean providing dial-up access for patrons who have terminals at home. In an academic setting this might mean putting terminals in all the libraries, in the stacks, in study areas and ultimately in dormitories and other non-classroom spaces.

Figure 13.9 Unobtrusive cabling in partitions. Photo courtesy of President Office Furniture

Maintenance Periodic equipment maintenance will require that access is available to the engineer or, more frequently, ease of access will be required to the printer for changes of paper and so on.

Construction material The material from which the workstation is made should be matt-finished (avoid chrome trimmings or shiny white surfaces) to reduce the risk of glare. (The table in Figure 12.1 (p.127) shows the effect of a shiny surface.) It should not be abrasive or subject to splintering, and it should be easy to clean and hardwearing. It should also be attractive. Opinions on what is attractive will, however, vary. Becket and Smith (1986) report a range of patron reactions to their reference desk which was made in an unusual wood. Comments ranged from 'it's gorgeous' to 'it goes well with your computers. It's got to be the ugliest piece of furniture I've ever seen in my life.'

The construction of the workstation needs to be firm, but heavyweight furniture may hinder flexible office arrangements. Items of furniture for particular functions should be well suited to their task. Printer tables, for example, should be stable, avoiding vibration when the printer is operating and should have adequate space for the storage of paper and the collection of printouts. All exposed corners should be rounded.

Computer security

Computer security and safety needs to be ensured, especially for microcomputers which are both desirable and portable. Minimal static electricity should be generated (see below) and a dedicated power line (that is, continuous from the service entrance) is recommended to minimize the likelihood of power surges. Air flow through the intake and exhaust vents should not be obstructed and large temperature and humidity variations should be avoided.

Static electricity

The potential for static electricity should be reduced. Static electricity can be generated by people and machines. People build up a store of static by creating friction with synthetic material, such as walking on synthetic carpets. The static is discharged in contact with metal furniture and a shock is received. Static can also be generated by, and transmitted to, the computer, causing potential damage to the software, the data and the computer itself. (For methods of reducing static, see page 67).

Effect on productivity

The demands placed on the furniture in order to create a comfortable and efficient workstation are greater than most ageing library furniture will tolerate. Although it is appreciated that financial restrictions will preclude many libraries from achieving the luxury of an 'ideal' workstation, it is worth noting the kind of furniture which is available, should funds permit.

Reference has already been made to unobtrusive cabling, adequate leg room, adjustable desks and suitable printer support. Although such ideals may seem remote, they are achievable using the specially designed system furniture offered by some of the suppliers listed in the appendix. Examples of such furniture are given in Figure 13.10.

Dressel and Francis (1987, p.1) investigated the value of improving the layout of conventional furniture of a group of employees (Improved-Conventional Group) compared with supplying furniture specifically designed for computerized workstations to another group (the Systems Furniture Group). They found that:

> The results of the study supported the hypothesis that improvements to the workstations would significantly, and positively, impact both productivity and satisfaction. The Systems Furniture Group demonstrated a significant increase in both productivity (20.6 per cent) and satisfaction . . ., while the Improved-Conventional Group showed a non-significant increase (4 per cent) in productivity but a significant increase in workstation satisfaction.

It can be concluded, therefore, that an increase in productivity is likely to result from an improvement in workstation design.

Figure 13.10 *An example of system furniture designed for computerized workstation.*
Photo courtesy of Project Office Furniture plc

14 WORKSTATION LAYOUTS FOR LIBRARY FUNCTIONS

Because of the variations in library operation, the following layouts for the issue counter, and workstations for cataloguing, serials control and online searching serve only as a guide to illustrate the theory which has been presented. They are by no means definitive.

Issue counter

The issue counter has many features which make it unique. Its special characteristics include: the need for staff to interact with users standing in front of them; a (normally) large number of books being handled; returns needing to be stored until they are re-shelved; the requirement for temporary storage for reserved items awaiting collection; frequency of movement; and, often, several staff behind the counter.

The aims of issue counter work have not changed, but automation has altered the way in which the task is performed and the shape of the workspace needed to support the operation. Instead of trays of cards, shared by all staff, there may be one or more data capture units. These may be specially designed units requiring to be built into the unit or they may take the form of VDUs and keyboards. Either way, change is needed.

Rea (1987, p. 18) describes the problems found at the University of Bath when VDUs were introduced to the existing issue desk:

> There was precious little room for the apparatus, and when Visual Display Units were introduced they had to be balanced precariously on the 9″ ledge. This was risky in itself, but additionally the rear of the unit with all its attendant wires was exposed to the public gaze creating a potential hazard. Also, some of the early models had their contrast control at the rear and this proved irresistible to some readers.

A different layout is therefore required for the automated system.

The primary consideration is perhaps the need for staff to be able to interact effectively with standing users. Factors which must be taken into account are the need to converse and the need to exchange items. If staff and user are too far apart, conversation can be strained, especially if the background noise level is high. De Jong and Roggeveen (1969) recommend a distance of about 100 cm between the two. The need to see the customer and to be able, easily, to transfer books implies that there should be no undue difference in working height between the customer and staff. Lifting weights up to or down from the raised ledge can be tiring, especially in a sitting position and can impose a strain on the spine. Rea (1987) reports that staff complained of back, neck and arm pains in performing these manoeuvres.

Grey and Wilson (1984) studied a number of issue desks and found that staff disliked too great a difference in height between the users' side of the counter and the staff side. Preference was given to counters of the following heights (Table 14.1)

Table 14.1 The difference in heights between staff and user sides of counters rated high by employees. Adapted from Grey and Wilson, 1984

Height of Counter		
Staff side (cm)	Borrower side (cm)	Difference in Height (cm)
95	95	–
90	110	20
89	106	17

Such counter heights on the staff side, make a tall chair obligatory. An adjustable draughtsman's chair is most appropriate, offering a swivel seat for ease of access and egress, castors and suitable back support. These are available from some of the suppliers listed in the Appendix. A raised chair also eases movement on and off the chair, if it is set at just below hip height. It is questionable, however, whether a raised counter on the borrower side is necessary. If a flat counter at a higher level is used, it will be within comfortable reach of the standing reader. It also offers the twin advantages of removing a now unnecessary barrier between staff and allowing more space for the VDU (Figure 14.1).

Wheelchair-bound users would also find a lower counter easier to use. This approach was adopted by Rea (1987) who settled for a level counter at 90 cm, the height also recommended by Grey and Wilson

Figure 14.1 *Flat top issue desk, suitable for a large automated library.* Photo courtesy of Terrapin International Ltd

(1984). Vertical panels were used to hide the back of the VDU and its associated wires and controls. These panels were dropped into slots which also housed the associated cabling (Figure 14.2).

Clarke (1985) proposes that the alterations should be even more radical with the member of public on the same side of the desks as the member of staff. He argues that people are happier with library systems when they can see what is happening on the screen.

Raised issue counters may help to solve some of the problems encountered in redesigning the workstation, but generate others. The trollies on which returned books are often placed ready for shelving may be too low for staff to reach comfortably from their raised positions, causing them to stand to return items and sit to issue. While this undoubtedly gives some of the movement which is necessary, it may also become tiring. Grey and Wilson (1984) found that shelves were preferred to trolleys for holding books. As these are height adjustable, they may be a preferred solution. Shelves for reserved items also need to be within easy reach of the operator. Usually they are situated at the back of the workspace and the operator must leave his chair. This is preferable to twisting to reach a heavy item.

The counter should allow enough space for all the staff. This means that allowance must be made for people to pass each other without

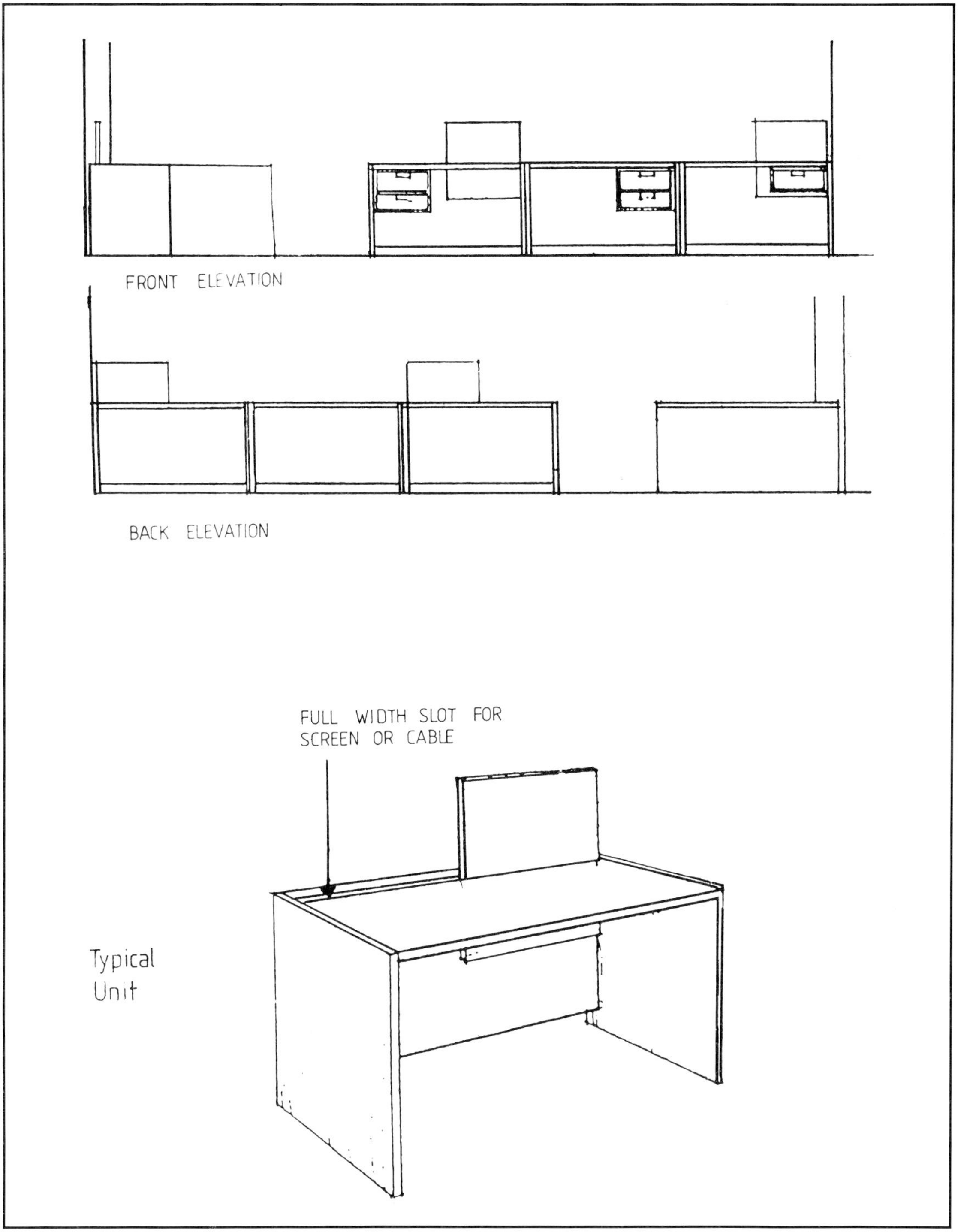

Figure 14.2 The issue desk designed for Bath University library. Source: Rea, 1987

having to 'breathe in', and equipment needs to be positioned in comfortable reach of all. The security device or cash register should not be situated for the convenience of one staff member at the expense of others, and the positioning of terminals at corners should be avoided. To ensure adequate working space, measurements can be made on the basis of the reach envelopes described in pages 139–142. Shared equipment may, if necessary, be situated in overlapping maximum reach envelopes but near-reach envelopes should not overlap. Grey and Wilson (1984) argue that psychological as well as physical space needs must be considered and a working width of 120–160 cm per person is suggested.

The depth of the counter needs to be sufficient to accommodate the VDU screen and keyboard (if used) but not so deep that the distance between staff and reader is sufficiently great (that is, greater than 100 cm) to impair communication. A depth of 70–90 cm is recommended.

Sufficient space needs to be created for the number of books handled at the counter and for any non-VDU tasks – for example, reader registration or alternative clerical work – which may be performed.

Cataloguing

Like all other library functions, the nature of cataloguing varies between libraries and even within the same library in terms of the demands it places on workstation design. Cataloguing may be retrospective or current. If retrospective, machine-readable records may be downloaded by online links to a MARC database or they may be input directly from the stock or the manual catalogue records. Current cataloguing varies similarly between searching for and downloading relevant records and creating original catalogue records. It is possible that a mixture of all these may occur at a single terminal.

The nature of the task will affect the way it is performed and the way in which the workstation must be designed to minimize physical discomfort and increase efficiency.

Cataloguing may therefore be divided into the following categories of VDU work.

- Intensive Data Input: for example, retrospective conversion from catalogue cards. This is usually performed by a typist copying from the cards and can induce high-speed input.
- Original Cataloguing: whether current or retrospective. This is frequently performed by a cataloguer, who may not be a trained typist and who may therefore need to see the keyboard as well as being able to manipulate the item being catalogued.
- Interactive or Conversational: where MARC records are downloaded from a central source online, for example. This may be done by a senior library assistant who inputs a key (for example, ISBN or author/file key), waits for the computer response, checks the screen for the accuracy of the retrieved item and selects the record for downloading.

The predominating task is the one which should dictate the layout of the VDU. In the last two cases, the following factors also need to be considered.

- Space is needed within the maximum reach envelope for a small pile of books awaiting processing.
- There must be room to open and move the item being processed (atlases and some picture books take up a lot of space and it may be necessary to write in the book – for example, to assign an accession number).
- Space is needed for a completed item. It is desirable for this to be close by and on the same level to avoid continuous lifting and twisting.

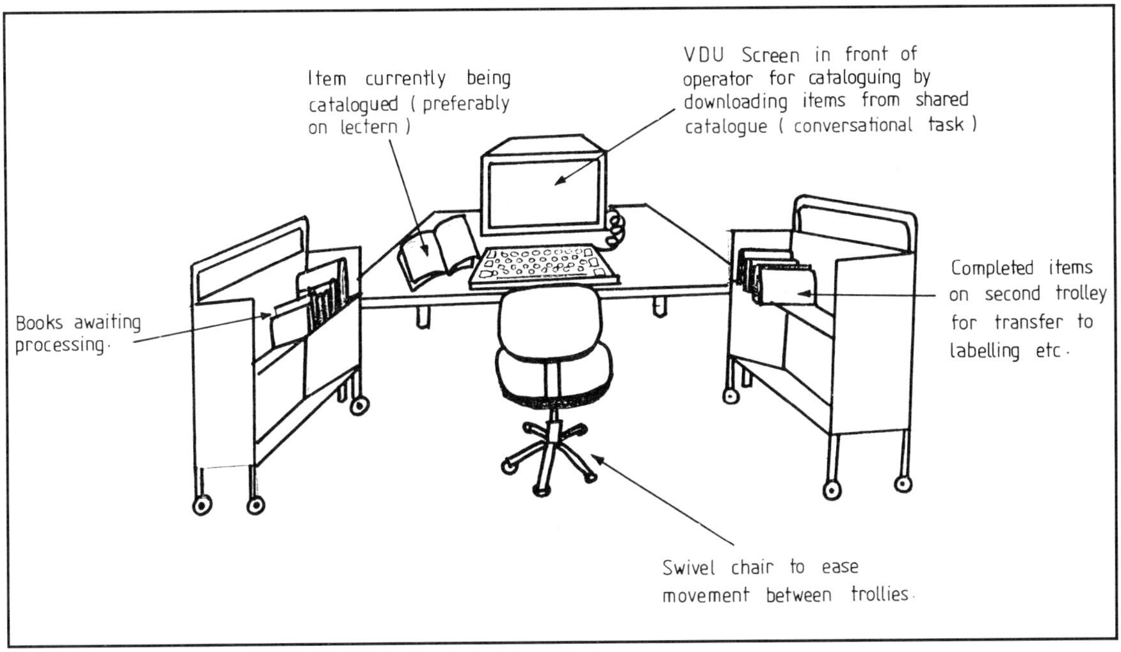

Figure 14.3 Suggested layout of cataloguing workstation for a conversational task

Serials control

Serials check-in is probably the most VDU-intensive part of the automated serials control process. The requirements of the workstation are: a buffer area for serials awaiting processing; a work area for the serial currently being checked-in with space to manipulate it to obtain details of volume, issue and so on; and an output buffer for items which have been processed. Additional space may be required for a printer to produce routing (circulation or allocation) slips if required and a date stamp to indicate the date of receipt. The operator will need to see easily the item currently being processed and to be able to check the data with that on the screen. Minimal keyboarding may be required. Some systems have now introduced a barcode reader for data input. However, as not all journals yet carry a barcode, a keyboard will still be required.

Figure 14.4 Suggested workstation layout for serials check-in

Online searching Online searching is essentially a conversational task requiring constant viewing of the screen and intermittent keyboard use. Printouts of search results need to be easily accessible and preferably within viewing distance during printing to assist in monitoring the search. Space is needed for the storage of manuals and thesauri, both those to which frequent reference is made and those which may only be used occasionally. Consideration also needs to be given to necessary interruption from and interaction with clients who may wish to be present at the search. Provision may need to be made for personal belongings and any alternative task which may have to be undertaken at the workstation.

Recommendations The above layouts are only suggestions for workstation designs as it is not possible to legislate for the range of library workstations which exist. In order to design the optimal workstation layout for a particular task, or set of tasks it is necessary to begin by undertaking a thorough task analysis. If other staff are affected, it is wise to get them to participate in designing the layout and to keep all the staff involved informed of your plans. Professional advice may be too expensive.

Once the analysis has been undertaken, it may be best to draw the proposed plans to scale on graph paper, to check that it is workable and to help solicit opinions from other staff if needed. It may even be helpful to set up a mock-up workstation with suitably sized cardboard boxes as these are both lighter and less fragile to move about than computer equipment.

As technology is continually changing, flexibility should be the keyword in choosing and arranging furniture and cabling, in order to cater for present and future needs.

Once the preferred workstation has been chosen, it should be left in use for a month or two, at which stage it is wise to check it over. Ask the opinions of any staff who use it with regard to ease of use, health and safety factors and general attitude. If problems exist, redo the task analysis and check to see if any improvements can be made.

It is in the nature of library and workstation design that no ideal solution will be found and compromises will inevitably have to be made. The objective is therefore to choose the layout which optimizes performance and maximizes operator well-being, comfort and safety.

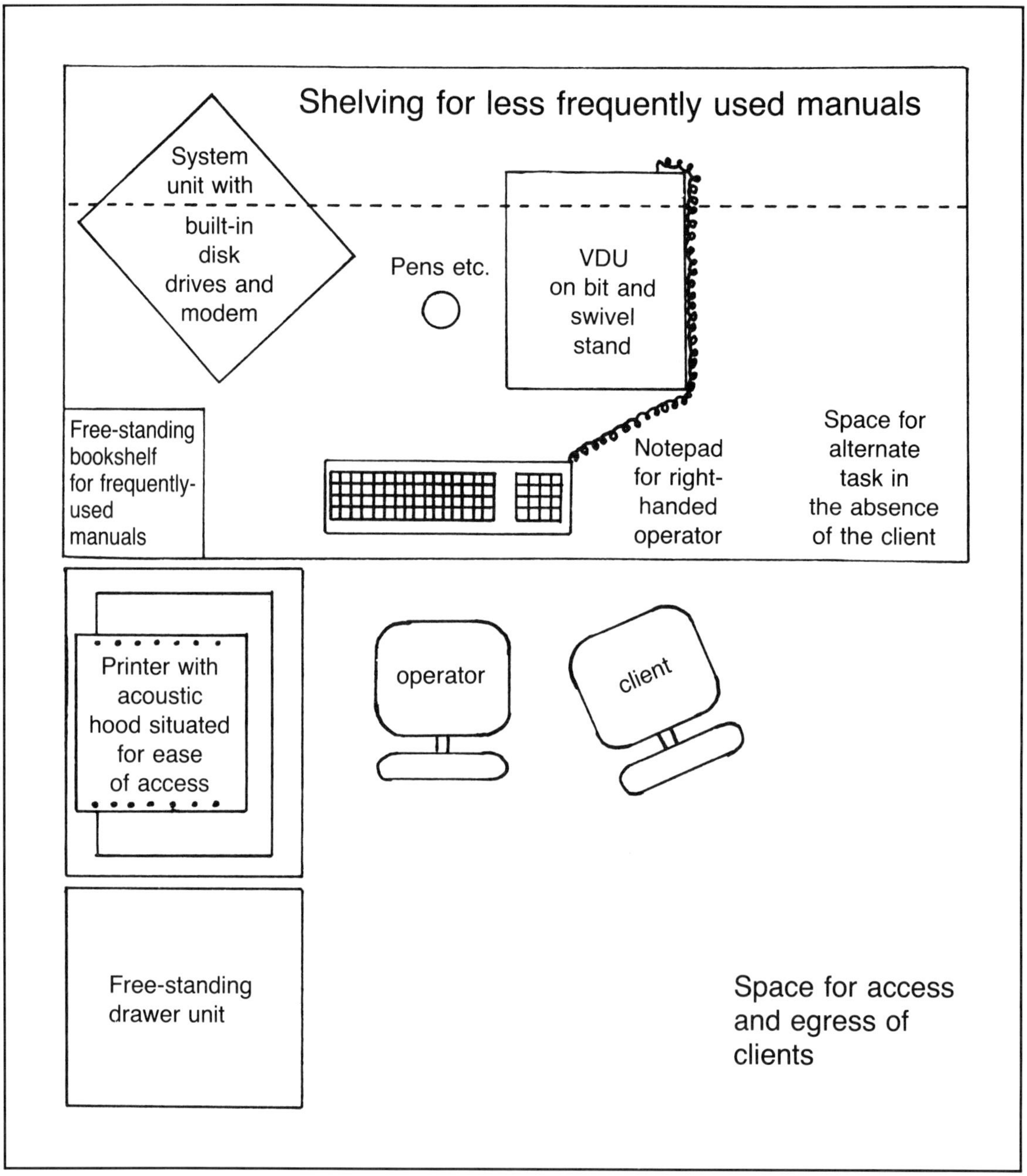

Figure 14.5 Suggested workstation layout for online searching

Part VI THE SOFTWARE INTERFACE

15 THE SOFTWARE INTERFACE

Introduction

It is well recognized that it is less cost-effective to commission tailor-made library software than to choose from the range of packages available. However, it is no easy task to choose from among the various options. Since the main reason for automating is to improve the operation of certain functions, it is important that the software package is able, first and foremost, to carry out the tasks required of it. Once a shortlist of suitable systems has been drawn up, it is necessary to decide between them. An important factor to consider at this stage is the human factor. Software which is difficult to use will be resisted by staff and will be used less effectively than a system with a well designed interface. If the objective of automating the library is to increase productivity, it is important to consider the influence of the software interface on operator performance.

The software chosen will need to be easy to use from a range of viewpoints: the operator, the library user, and the task to be performed, and a number of other factors will affect its ease of use.

User requirements

In order for a software package to be an effective tool, the user needs to be able to interact with it without frustration, boredom or confusion. This means that the dialogue used by the system should be appropriate to the user's level of skills and knowledge. (The term 'dialogue' is used to mean the instructions that the user issues to the computer and the information that the computer displays for the user.)

Types of user

Computer users are not uniform in their experience, skills or expectations. Cole *et al.* (1985) identify four groups of users of computer systems:

- *Naive users* who have little or no knowledge of the task to be performed or of computer systems. OPAC users will include a large number of naive users.

- *Casual experts* who understand the application thoroughly but have little computer knowledge. A librarian, well versed in cataloguing, for example, but used to manual systems only, would be a casual expert.
- *Associative experts* who have considerable familiarity with computer systems but little knowledge of the particular application. A possible example would be OPAC users who are students from a computing studies course but who have little library experience.
- *Experienced professionals* who have skills in both computing and the task for which the computer is used.

Cole and his colleagues point out, however, that these categories are not discrete and that people move between them. A novice user of an OPAC, for example, may become a casual expert in time or an experienced 'pro' may regress to a casual or an associate expert through lack of system use.

The type of dialogue best suited to each category of user varies with the user, the task performed and the frequency and duration with which the task is performed. A novice user, for example, needs prompts indicating what options are available and the implications of selecting a particular option. A casual expert requires to be guided through the system, but can comprehend terminology relating specifically to the task. The converse is true of the associative expert who is unlikely to need to be prompted to hit the RETURN key but may require information on the function being selected. The experienced 'pro' will want power and flexibility above all and could become irritated by the distractions of prompts and help messages.

Thus, in choosing software, the nature of the predominant users must be considered. Complications arise, however, because of the changing nature of users. Users not only change categories over time, they may change according to the task being performed, so that an experienced 'pro' online searcher may become a mere associative expert when operating the issue system for example.

One solution may be to acquire software with an adaptable interface. Some software, for example, offers a default menu but with a 'hidden' command-driven interface for experienced users.

User characteristics (see also Chapter 1)

All people have characteristics which affect the way in which they process information which, in turn, has implications for the software interface.

Memory

As discussed in Chapter 1, memory is divided into short- and long-term memory. Short-term memory can be disrupted by the receipt of competitive information, such as an interruption by a colleague or the telephone. Some systems demand the use of short-term memory (or a notepad!); for example on some issue systems if a reader number is not known it may be found by searching the reader file, but it must then be input manually in order to issue an item. If interruptions are anticipated as part of a task, it is desirable if minimal use is made of short-term memory.

Long-term memory is less prone to information loss, but items can become temporarily unavailable (Lansdale, 1985). Therefore command words may need to be available on help screens or menus so that they are accessible to authorized users, but need not be permanently displayed.

Mental capacity

People vary in their ability to receive and process information (see, again, Chapter 1). Some people become bored or frustrated with too little information, such as having to scan through several screens each of which contains only two or three items, and others can suffer from mental overload if given too much information. A poorly designed, cluttered screen, for example, may bombard the user with information causing him to miss salient items. Seton-Watson, a Scicon consultant, in a criticism of software use in Defence, argued that 'human operators [could not] increase their abilities to process information beyond fixed limits.' ('Defence system designers urged to improve displays', 1985).

Perception

The ability to recognize an individual item of information from a crowded screen is affected by the user's experience and expectation and the way in which the item is displayed. An operator who is experienced in the use of the system and is familiar with the data and their position on the screen will be quicker to perceive the relevant information than a novice user. To assist a novice user, therefore, the system needs to use various facilities to highlight specific items to attract the user's attention. Commonly used stimuli include bright colours, bold or flashing text and inverse video. Bleeps are also effective.

Barriers to learning

Carroll and Rosson (1987, p. 80) identified that some 'people have considerable trouble learning to use computers'. The problem, they found, was that many people are motivated by productivity rather than learning. Initially, therefore, their response is to 'jump in' and try out certain procedures in preference to reading a manual or screen-based instruction. When in doubt, then they may turn to a source of advice to find out where they have gone wrong. There are times when this kind of user will persist in incorrect or inefficient methods of performing certain tasks because, as they achieve the desired result and there is no apparent error, they will not be motivated to seek help.

Carroll and Rosson also identified that the skill of such users tends to peak at relative mediocrity. Once again, learning time is seen by the users to be ineffective, so they will adhere to known procedures to achieve a result, rather than proceed to acquire greater skills.

Task requirements

In order for software to be effective, it must be capable of performing the task(s) required of it. This demands that a thorough analysis of the task be undertaken before the software is considered. (Guidance on how to select software to perform particular functions is available elsewhere in the literature.)

From a human factors viewpoint, however, the way in which the system achieves the function is just as important as what it achieves. A system which is cumbersome in use will offer few benefits for the operator (if any) over a manual system, even if advantages such as increased management information statistics are available. The frequency and the order in which certain operations must be performed should be mirrored as far as possible by the software interface. For example, the way in which information is arranged on the screen should concur, as far as possible, with the order in which this information will be used within the task and, preferably, should be compatible with any documents that are used (Sperandio, 1987). That is not to say, however, that operations should not be changed, but rather that due consideration should be given to those which cannot.

Interface design

The term 'user-friendly' is often used to describe software which is easy to use, but it is seldom clearly defined and is often somewhat subjective. There is a distinction between a system which is easy to use for a novice or one which is easy to learn, and software which is suitable for the experienced 'pro'. The former usually offers choices or prompts, the latter relying more on commands issued by the user. The ease of use of each interface may be identical, but for different users. Galitz (1984) defines the qualities of easy-to-use software as: adaptive; transparent; comprehensible; natural (that is, it mimics the worker's behavioural patterns); predictable; responsive; self-explanatory; forgiving; efficient; flexible; and available (that is, 'down'). He also lists a more objective and measurable set of features which are explained at greater length below:

Consistency

A system is easier to learn and to use if it is consistent in its use of commands, its displays and its behaviour. Consistency of position of prompts and error messages will increase user expectancy and increase speed and accuracy of operation. Consistent use of commands and function keys is also important. One library system, for example, is inconsistent in its use of the RETURN key, sometimes requiring its use and at other times not. Consequently, errors are frequent and users become frustrated. Consistent use of function keys, too, to move forward, escape or save, reduce the possibility of user error and speed the learning process.

Initiative

The degree of initiative taken by the software or expected from the user will vary. Software which takes the initiative is like the Prestel interface where choices are offered by the software and selected by the user. An example of software which leaves the initiative to the user is that used on online hosts, where there is a wide range of commands which can be issued by the user at any time and in any order. The type of user who will be operating the system will dictate the amount of user initiative which is desirable. A novice will require the computer to prompt and give choices from which selections can be made, whereas the experienced operator will probably prefer to be able to choose both the task and the pace at which he will work.

Information load

A person has a limit to the amount of information he can process – that is, his 'mental load'. If the load is too heavy, stress may result, if too light, boredom may occur. A software interface can add to an operator's mental load by demanding the use of a complex command language or by using complex screen designs. In contrast, an operator may become quickly bored by having to repeat several simple commands, rather than being able to input a single more powerful instruction, or by the need to page through several screens to locate a required item. Frequency of use of the system, motivation and other user characteristics will affect the degree to which an operator will be stimulated or bored by a particular system.

Control

The degree of control which an operator has of the program will also affect his feelings towards the software package. The user should always be in control of the interaction. A job which is paced by the machine, or which in any other way makes the user feel the machine is in control, can create stress.

Dialogue style

There are a number of different ways in which commands may be given to, or received from, the computer system and they may be appropriate to different categories of user. The more commonly found examples are detailed below.

Batch input

Most software, nowadays, offers online files' updating but, on some systems, input of large quantities of data may still be required in batch. In addition, some programs offer both online and batch facilities. Online updating of files means that changes to the file are used to update it as they are input, record by record. This has many advantages, ensuring currency of data and immediate access to updated records from elsewhere in the system. To input by batch means that data are entered into a holding file and then used to update the system files in a single process. Batch input has some advantages over online input: it may be faster, because time is not lost waiting for each item to be individually added to the system, and word processor editing facilities are often far superior to those offered by the data input function. Its main disadvantages are that there is a delay between typing in an entry and its input to the system, and input must usually be clearly identified and in a predefined order. Consideration therefore needs to be given to the type of task for which each form of input is most appropriate.

Prompted input

The input function on some systems offers a series of prompts for data input. Data must be entered or the RETURN key pressed in response to each tag in order to pass on to the next prompt. This form of dialogue offers the advantage of guiding the user, but can prove to be rather rigid in use; movement up and down the screen to edit items can be restricted.

Form filling

Frequently, catalogue input or database programs display a form on the screen and individual items of information are entered in each section of it. This offers the twin advantages of prompts for the user and an overview of the whole record. In order for form-filling to be easy to use, full-screen editing is required to enable the operator to move between specified fields. Ideally, too, the order of input should be from left to right and from top to bottom. If data are to be input from a worksheet, it is also desirable for fields to be displayed in the same order.

A form layout may imply a restriction on the amount of data which can be input to certain fields. In practice, this limitation may be only on the amount of data which may be viewed at any one time, additional information being scrolled horizontally or vertically out of sight. While it is possible to input large quantities of text, the difficulty of editing it must be considered.

Menu options

Many systems, especially OPAC, offer the user a series of choices from which a selection may be made. This can be particularly helpful to the naive user who would not otherwise be aware of the options available. However, in a frequently used system the mandatory completion of menus may become frustrating. Some systems offer an alternative command system or enable users to input a string of menu selections which take them directly to a function without displaying the menu. Hidden menus are sometimes also available as a prompt to less frequent users. These may be available at the press of a function key, or there may be an option continuously to display or hide them. Man's decision-making process (see page 15) means that people tend to place much importance on the information collected early in the process. Thus in a menu system, the most frequently used and most important items need to be placed towards the top of the menu or they are likely to be ignored. Man's capacity to remember unrelated items also has implications for the length of the menu (see page 10). Long menus tend to confuse people, and it is recommended that the number of options in a menu should be less than or equal to seven and should never exceed nine.

Command languages

Command languages, as the name suggests, are a vocabulary of instructions which the user can issue to the computer system. There are several features of a command language which affect its ease of use.

The words used should be readily understood by the operator. Most of the command languages used in automated library systems are designed by computer programmers rather than by librarians and may not always reflect the terminology with which the latter are familiar.

Lansdown (1984) examined users of text editing systems and found that there was little agreement on the command words that people preferred, but that where there was agreement, the following words were preferred:

Add for insert
Omit for delete
Put for move
Switch for transpose.

This list also highlights the transatlantic differences in terminology which exist. Similar arguments may exist over library terminology, for example:

Check-out or issue or lend or loan or charge
Check-in or return or discharge.

The command words should be both easy to remember and easy to distinguish from one another and from other commands that the operator may use on other systems. It is desirable that commands can be shortened, but any abbreviated command should remain memorable and should not easily be confused with others. One system, for example, uses the abbreviated command ^KD (Control-KD) to finish working and to save the file, while another system uses FD to leave the system and to delete the file. Therefore, while an interface may be consistent within itself, it may be dangerously confusing to an operator using more than one system.

In addition to the commands issued by the operator to the system, the program will issue instructions to the user. The system commands given will convey an impression to the user. The National Electronics Council (1983) cites the following command: 'You have been idle too long . . . Please respond . . .' (Yes, sir!). While people who are comfortable with computers may simply ignore such prompt, it could be off-putting for new or timid users (see also subsection 'Errors' below).

WIMPs

This acronym stands for Windows, Icons, Mice and Pull-down menus. Some people refer to WIMPs interfaces as window or iconic interfaces. The objective of this interface is to offer the power of a command-driven system with the light memory load of a menu-driven system. Mice (see subsection 'Mice', p.103) are used to control the cursor and to select commands from the menus (see Figure 15.1). Menus and sub-menus usually appear in blocks of text or windows which temporarily overlay part of the screen. Icons are used to indicate graphically the function which is to be performed – for example, a dustbin is sometimes used to indicate that a file or record may be deleted. This interface is common on some personal computers, such as the Apple Macintosh, but has not yet reached the mainframe environment, nor indeed many library systems.

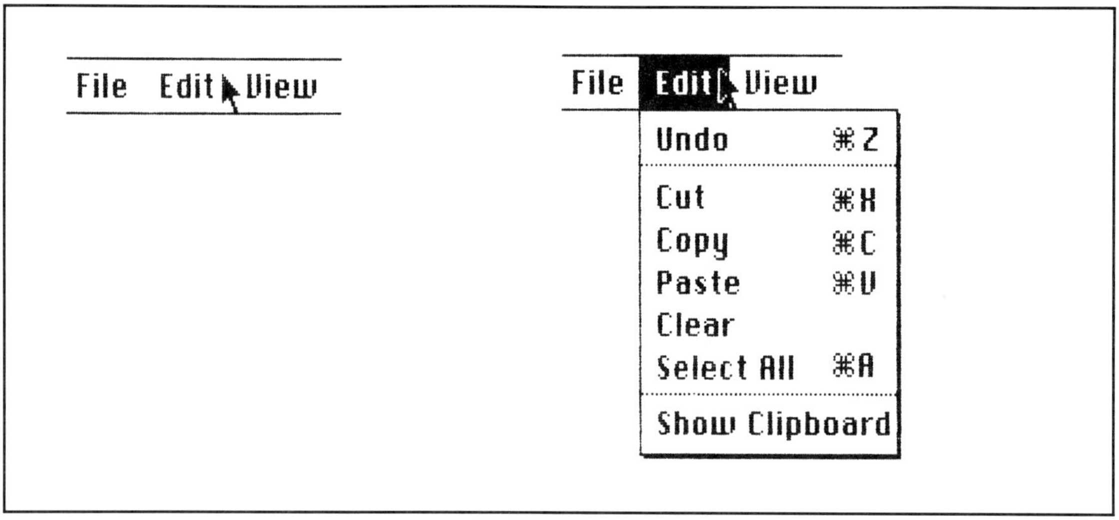

(a) Cursor positioned to select EDIT menu **(b) A window displaying the EDIT menu**

Figure 15.1 An example of the use of a pull-down menu

Which is the best dialogue style to use?

No single dialogue can be pinpointed as the most suitable, rather a variety of dialogue styles is desirable to serve the differing purposes of task and user. For example, the contents of a database may be entered using a form-filling dialogue, and may be interrogated by an experienced 'pro' using a command-driven interface or by the naive user on a menu-drive OPAC. It is rather simplistic, however, to assume that novices always want menus and that experts prefer command languages. Antin (1988) found, in a comparison between menu selection, command entry and combined modes of computer control that, although command entry produced superior performance to either a menu or a combined interface, operator preference was for a combination of commands and hidden or pull-down menus. Interestingly, the operators felt that their performance was faster using the combined interface, although the results showed that use of the command-driven interface had produced the fastest times.

Help

The level of help required will depend upon the individual user. Initially, a user will require clear prompts and other forms of system guidance. As expertise grows, however, it is desirable to be able to hide the help so that it may be called up only when needed to serve as a reminder or to guide the user through a new routine. Help messages should be easily understood and should avoid jargon. Many users may need only a reminder of the set of commands which can be issued at a certain point so there should be options within the help system for more or less detail as required. Help screens are not intended as a substitute for the manual and should refer to it if long and complex explanations are needed.

Feedback

Nothing is more irritating than a system which does not inform the user of progress. If a file is being accessed, or an item has just been updated it is helpful if the system displays a message to that effect, especially if there is a delay of more than 10-15 seconds (Galitz, 1984).

Recovery

It is desirable to be able to retract a command. For example, in response to the command to delete an item, many systems offer a second chance. 'Are you sure [Y/N]?'. Similarly, in some word processing systems it is possible to recover the last item deleted.

Another form of recovery is the ability to quit from any point in the system. This is particularly important for those users who like to learn by exploring the system and who may then get 'stuck'. One text retrieval system, for example, uses an asterisk, to enable the user to exit from any function in the system.

Funcation keys

The use of function keys can reduce the number of keystrokes and the memory load of the operator. In order to access some online systems, for example, it is necessary to key in long codes consisting of unrelated alphanumeric characters. Use of a single-function key which automatically inputs the code can improve speed and accuracy. The keys need to be clearly labelled, consistent in their use and arranged in logical groups. Failure to do so can result in errors.

Errors

However good the user, some mistakes are bound to occur, and these will vary in type and frequency. Borgman (1988), in referring to the use of information retrieval systems, divides errors into the two main types: those which are due to the mechanical aspects of searching; and those which are conceptual in nature (that is, due to a failure to understand search logic). Mechanical errors may be of two kinds: logical errors, such as an incorrect command sequence; and typing or spelling errors. Reporting a study by Dickson (1984), Borgman claims that 15 per cent of the errors in title searches could be attributed to typing or spelling errors, the remaining errors being conceptual in nature. Such a high percentage of errors caused by an inadequate understanding of the system, points to a need to establish good training and support procedures. As experience grows and training is undertaken, the ratio will undoubtedly change, more of the errors being due to typing or spelling inaccuracies, and the overall number of errors should fall.

The help given by the system in response to user errors is also important. Buxton and Trenner (1987) in their study of error messages in information retrieval systems found that even those systems which claimed to be user-friendly, tended to respond with hostile error messages. They argue that error messages need to be polite, specific, constructive and helpful and not use 'hostile' features such as cryptic codes, cryptic vocabulary ('Storpstg ovfl'!), dominant commands (such as 'LEFT HAND TRUNCATION IS NOT ALLOWED. REENTER THE QUERY CORRECTLY') or emotive vocabulary. The tone and content of error messages, they believe, can significantly alter users' behaviour, sometimes heightening their anxiety. They cite a paper by Shneiderman which concludes that 'improving messages can play a significant role in improving user performance and attitude'.

Response time

The system's speed of response should be compatible with the task to be performed, but will not be dictated solely by the software. A task requiring continuous data entry – for example, from catalogue cards – requires a rapid response between records, whereas a longer delay may be acceptable when inputting from a book or journal as time will be taken in handling the items. If the operator is free to go and do other things, speed of response is not important.

If response time is too fast, however, the operator may begin to feel paced by the computer. During data input, particularly, a fast response may encourage intensive, repetitive work, with its attendant problems (see pages 28–31).

A variable speed of response can also cause frustration and stress. This is particularly likely to occur on shared mainframe or large multi-user systems when large jobs, such as calculating fines or preparing and printing orders, slow the system down from time to time. Where the variable response is due to the difficulty of the task in hand, frustration is less likely to arise; operators tend to exercise more patience when they know that the computer 'has a lot of work to do'.

Screen design

Screens may be used to capture or to display information. The screen display has three main functions:

- To display information.
- To give the system status. It should always be clear to the user where he is in the system.
- To show the potential inputs (what information goes where or where can he go next?).

The design and layout of the screen will affect the ease and speed with which these functions are fulfilled. Good or poor screen design may be brought about by a combination of the following features:

- the information content of the screen – that is, the data, prompts, messages and so on;
- the formatting of the information; the use of windows, grouping and so on;
- the coding of the information, the use of colour and so on.

Information content

The information displayed on the screen should be both consistent and relevant. Consistency, for example, should be maintained between different screens or functions. A menu system which lists an option as 'search and catalogue' should display the same phrase as the heading for the catalogue search screen, rather than using a briefer but different term, such as 'catalogue enquiry'. To the initiated, they may appear to convey the same information, but to the new or uncertain user they can cause confusion.

The relevance of the information may depend on the user or on the task. Some smaller systems try to combine a lot of information on a single screen (see Figure 15.2). It is unlikely, however, that all the information will be required for a single task with the result that, whatever task is being performed, the user will have to scan a crowded screen to locate the relevant information, with subsequently reduced efficiency (see 'Formatting' below). Additional information which may be required should be available in pull-down menus, rather than cluttering the screen.

```
ITEM NO 81972,          CLASS 121-A378-B   ITEM TYPE D10   PRINT (Y/N)   Y
                        AUTOMATION IN LIBRARIES
                        ____________________________________________
                        ____________________________________________
SR/CONV                 COMPUTERS; LIBRARIES
AUTHOR                  BROWN, F. __________________________________
                        ____________________________________________
                        ____________________________________________
COMMON AUTHOR           N/A ________________________________________
PUBLISHER               BLACKWELL ____________ CITY OXFORD __________
EDITION 2 VOL/ISS ___________ ( ___________ ) PAGES 560 ILL. N PUB. YEAR 1984
STATUS 7     LEND.STATUS 1     LEND.PERIOD 10                     1   2   3
CONFER 18277     LANG. CODE E      STOCK (ORIGIN)                24   2   2
LENDINGS 256            STOCK (SHELF)                            19   2   2
CIRCULATION LIST 127    RENEWAL DATE ____________________   FREQUENCY
REMARKS EXTENDED LENDING LOCATION UNIVERSITY BUSINESS SCHOOL
ISBN 0 631 14610 5     INDUSTRIAL RELATIONS
READERS   1926   1   274   3   20911   2   7081   5   999   4
```

Figure 15.2 A combined screen for catalogue and issue data. Too much information on a single screen can appear cluttered and reduce efficient operation.

It is important that, at all times, there is an obvious indication of what is being shown and what should be done with it. It is also important that the user is given a simple way of finding what is in the system and how to get it out and a clear indication of any action which could make a permanent change in the data or system operation.

Formatting

The layout of the data on the screen will have a considerable impact on the effective use of the data. The following criteria may be helpful in determining what is or is not a good display:

- The screen should have an orderly appearance and should not be overcrowded. A cluttered screen makes it difficult to locate specific information, causing an increase in search times and errors (Galitz, 1984). Galitz (citing Dunsmore, 1982) reports on a system where separate items, which had been combined on the same display line to conserve space, were instead placed on separate lines, with the result that users of the new screens were 20 per cent more productive.
- Consistency of layout is important so that, if the user becomes familiar with a certain prompt or error message appearing in a particular place, his performance will increase as his visual search times are reduced.
- A logical ordering of data is also important. The form for catalogue input, for example, should be in the same order as any hard-copy form which is used, or should be arranged to permit a logical and orderly input of data. System prompts should also be placed in positions which seem logical to the user. For example, instructions on how to use the screen or its information should be at the top and instructions on what to do with the completed screen should be at the bottom. Other logical groupings are: by sequence of use; by function; by priority or importance; by frequency of use (Cole *et al.*, 1985); or alphabetical or chronological orders.

```
NEXT [2      ]  SCREEN  [1CAT]  TASK [EDIT      ]            REF.            2

                MAIN CATALOGUE RECORD DETAILS — INPUT AND EDIT

  Title         [Introductory Experiments in Digital       ]   [intrexpe]
                [Electronics and 8080A Microcomputer       ]
                [Programming and Interfacing ( Book 2)     ]

  Author        [Larsen, D. G.                             ]   [larsdg  ]
                [Rony, P. R.                               ]   [ronypr  ]
                [Titus, J. A.                              ]   [tituja  ]


  Publisher     [Howard W. Saas             ]
  Location      [London              ]  Year of publication  [1979]
  Collation     [                 245PP]
  ISBN          [  0-241-10581-1]   ISSN [          ]  Class  [ 681.32.6]

  Notes         [                                                   ]
                [                                                   ]
                [                                                   ]
                [                                                   ]
```

Figure 15.3 A catalogue input and edit screen. The grouping together of related information and separation of unrelated data speed the location of data and increase operator performance

- Groupings of information can improve retrieval speed. Related data should always be grouped together and clearly separated from other groups by space and position. It is undesirable for related data to be spread over separate screens. The various elements (headings, prompt lines and so on) should be clearly distinguishable by their size, colour or position. The partitioning of data should be consistent and, whilst being clear, should not distract the operator. Large numbers of similar-sized partitions should be avoided. The display in Figure 15.3 shows good grouping of data into logical sections: title; author; imprint; reference numbers; and notes.

Coding

Screen information may be coded for a number of reasons; for example, in a free text system, a word or phrase may be highlighted to indicate what caused the record to be retrieved, or a defaulting borrower may cause the issue system to bleep when he next borrows or returns an item. A variety of coding techniques may be used, including: position; brightness; flashing; inverse video; colour; size; and sound. All of these are used to draw the operator's attention to an item. However, because of the way in which people respond to external stimuli (see Chapter 1), some codes may be more effective than others. A continuous beeping sound for example, is more likely to draw attention than a single beep (and will be more annoying!), and a combined visual and tonal stimulus will be more effective than either in isolation. Too many different stimuli, however, will be rendered ineffective, because it is not possible for the operator to concentrate on several items simultaneously.

The use of colour on screens is becoming increasingly common and, with it, some examples of the poor use of colour. Galitz (1984) warns that the use of colour may cause a person to notice differences in colour regardless of whether they have task-related meanings or will tend to group together items of the same colour in a way which is unrelated to the task. He argues that, when colour is used improperly, performance will show little improvement and may even suffer. He also points out that about 8 per cent of the male population and about 0.4 per cent of women have some form of colour vision deficiency. Nevertheless, despite the drawbacks listed, colour can be used as an effective coding device in a variety of ways.

Groups of related data may be tied together by using similar colours, or unrelated groups of information may be coloured differently to separate them. Spatially separated related data may also be linked by colour. The use of colours to group or discriminate will depend on their position in the visible spectrum (Table 15.1). Colours which are close in the spectrum should be used to link items together and those which are spread out along the spectrum should be used to distinguish items.

Colour may also be used to draw attention to important fields. One database system, for example, uses colour to draw attention to the terms within a field which have been indexed. In order to attract attention, it is important that bright colours are used. The relative brightness of colours, in decreasing order, is white, yellow, green, blue, red (Galitz, 1984), or of those used in videotext displays, white, yellow, cyan, green, magenta, red and blue (Van Nes, 1986). Of these, the most easily visible colours under a range of lighting conditions are green and yellow. The human eyes perceive blue most easily in poor lighting and red in bright conditions (see page 35). The lighting conditions surrounding the VDU display will therefore affect the ease with which colour coding is seen. It should also be noted that if blue and red are displayed together, under normal lighting conditions, red will dominate.

Table 15.1 The visible spectrum of colour

Colour	Approx. wavelength in millimicrons
Red	700
Orange	600
Yellow	570
Yellow-green	535
Green	500
Blue-green	493
Blue	470
Violet	400

Source: Galitz, 1984

Colours may also be used to assist the reading of long lines of text or data. This may be done by using three row groupings of alternate colours to guide the eye when moving from one line to another (Galitz, 1984).

The aesthetics of colour coordination should not be forgotten. Too many colours or the use of uncoordinated colours can make a screen confusing or unpleasant to look at. Galitz (1984) recommends that alphanumeric screens should display no more than four colours at one time. The choice of colours, however, will be a matter of taste and will vary from person to person.

The emotive effects of colour are also important. Red, for example, conveys the concept of a warning or of accounts which are in difficulties.

Practical advice

There are many factors which make software easy or difficult to use and in order to choose between competing packages, the following course of action is suggested:

- See demonstrations of software and, if screens can be redesigned, ask to see the design you want prior to purchase. Do not assume it can be done.

- Use the system and check it in relation to the above criteria.
- Visit other users of the system and assess the reactions of the *operators* (rather than the managers) concerning its ease of use.
- Allow the potential operators to see and use the system and obtain their opinion, but be aware that if they are unsure of working with computers they may not be in a position to judge, and use of a system without proper training may repel, rather than attract, them.

Part VII JOB DESIGN

16 JOB DESIGN

Introduction

The major aim of staff training is to make the library staff more effective so that they can offer an improved service to readers. It is not enough, however, that staff are taught to operate equipment and undertake necessary routines, they must also have the motivation to ensure the smooth running of the service. Well-motivated employees have the potential to contribute beyond the narrow specifications of their jobs.

Motivating staff is a complex and sensitive issue. There are many variables determining an individual's sense of job satisfaction. His own attitudes play a part, as do the prevailing management style, the nature of the work group, and the nature of the task. An organization cannot afford to disregard motivational theories entirely or poor morale (leading to haphazard performance and general indifference) will result. The introduction of an automated library system is therefore a unique opportunity to reappraise work activities and conditions. Since the whole concept of technology revolves around changing the way in which tasks have been done, it is not sufficient to superimpose one system on another: due attention must be paid to the consequences as they affect job design.

The implications of information technology will vary from area to area of library work according to what is being installed. Nevertheless, computerization will impinge everywhere in more or less subtle ways. Some jobs will disappear, others will be created, merged or expanded: responsibilities will alter. Automation will have contrasting effects on the tasks of professional and non-professional staff and its impact can be construed as either positive or negative. Neither must its effect on the working environment be underestimated. In the following analysis the focus is primarily on where job satisfaction may be damaged by the introduction of computers. The overall worth of the technological initiative to the organization should have been established in the preliminary stages of the project and concentration should be given to making it as successful as possible.

Job design and job satisfaction

Job design is important from the twin standpoints of the organization and the individual. Failure on the part of management to devise satisfying and complete roles for their employees can create intolerable levels of strain and stress, which in turn can lead to a high rate of absenteeism and labour turnover. Informing staff that their jobs are structured in order to improve efficiency will not in itself guarantee productivity; it is better actively to involve employees in the quest for increased effectiveness and thus promote team effort.

Principles of good job design

There is a vast literature on the topic of what constitutes a 'good' job and the many factors which influence people's criteria once their basic requirements of security, reasonable working conditions and adequate pay have been met. Starting from this documentation and other personal studies, Sell (1986) – of the Work Research Unit – has compiled a list of principles on which to base sound job design: variety; autonomy; feedback; social contact; identity; responsibility; achievement; opportunities to learn and develop; role under/overload; role conflict; role ambiguity; to which should be added the need for adequate rest breaks. These are discussed in more detail below.

Variety

It is desirable that staff should carry out different tasks at different speeds and in different places. The clerk at the issue counter, for example, should be offered other sub-tasks and ample opportunity to work away from the counter for periods.

Routine jobs are likely to induce feelings of extreme fatigue. It has frequently been promised that technology will eliminate dull and repetitive jobs. Malinconico (1983) remarks that this is partly true but that they are replaced to a large extent by equally unstimulating tasks, an observation which may be true of libraries. The Library and Information Technology Centre (1987, p. 21) states that many libraries 'claim a substantial increase in productivity as a result of more efficient circulation systems': what this has meant in practice is more shelving! Sophisticated technology has, in some cases, paradoxically promoted the least glamorous and most derided chore of librarianship.

Autonomy

Staff should have some say in how they carry out their work, such as the order of the sub-tasks. It is important that use of an automated system does not impose too rigid a method of working.

Feedback

The job-holders should know how well they are doing without having to wait for feedback from management. Achieving a clear desk at the end of the day can be enough in itself but, if backlogs develop, it may be necessary to set a realistic target which, when reached, indicates to the operator that he is performing satisfactorily. Computerization also reduces the *visible* evidence of a task well done. At the end of the day, there is no physical manifestation of expended effort. Statistical data confirming that so many transactions have been completed do not perhaps generate the same degree of satisfaction.

Social contact (see also 'Employee–employer relationship' p. 52)

This, for most people, is a desirable part of any job. While it is appreciated that time must not be wasted by unproductive chatter, it must also be recognized that the opportunity to see and converse with others can be beneficial to staff morale. Workstation layout should enable social contact but minimize unnecessary interruption.

One area where automation has an obvious impact on social contact is the issue desk. Staff have commented that the time taken to issue or discharge a book using the Browne issue system provided an opportunity for social interaction with the user. The reduced time required by most automated systems renders such contact impractical. The use of OPAC can also reduce opportunities for communication, especially where users are able to place reservations automatically.

The implications of VDU usage are, however, far wider than just the library context. Electronic communication has revolutionary potential for society. Smith (1981) delineates the scenario in which private workstations at home or in the car will be the norm. Group teleconferences, data and text transmission, all theoretically allow work to continue independently of geographical location. People are no longer restricted by the physical transfer of material. Many academic libraries are linked to campus networks and distance learning has become a reality. Weise and Borgendale (1986) give an account of the Electronic Access to Reference Service (EARS) operated by the Health Sciences Library of the University of Maryland and Baltimore. Not only can online catalogues be reached from remote sites but other library services can be provided electronically. Specific requests for information, photocopies and interlibrary loan orders can readily be transmitted without the user setting foot in the library. Interestingly, it is noted that while EARS further reduced external communication, it increased *internal* cooperation between library departments who have to resolve difficulties created by an invisible public.

Identity

Tasks should fit together to form a coherent job. At the issue desk, for example, it is desirable for the operator to be able to deal with all of the reader's requests rather than having to refer them to someone else to alter their record or to make a reservation. Pearce (1986a, p. 1) comments: 'Too often jobs are not designed, but happen by default: they are the tasks left over that are inconvenient or impossible for the computer to perform.'

Responsibility

Having a coherent job enables the user to be given responsibility for it – for example, an assistant may be made responsible for checking the stock and tidying the books on a particular stretch of shelving or of organizing displays.

Responsibility and the need for certain skills can be reduced by automation of some functions. If a shared cataloguing system is adopted, for example, the cataloguer may spend most of his time downloading matching records rather than creating new ones. He may be concerned about the reduced level of responsibility or the underutilization of his skills. Similarly, German *et al.* (1987) highlight that online searching, although a professional task, demands a number of record-keeping activities which, although clerical in nature, fall to the search professional. It is worth noting that an operator whose job has been deskilled will experience more stress than one who has always performed the less skilled task (Cakir *et al.*, 1980). Rooks (1988) observes, too, that personal authority over information will diminish. To determine whether a book is on order no longer means consulting the acquisitions staff, even the reader can find the information via a public access terminal.

Increased responsibility may also result from the introduction of an automated system. The task of searching and downloading records from a shared catalogue may be delegated to a senior library assistant, while the professional cataloguer deals with complex items and with the supervision of the clerical staff. This may enhance the work and job satisfaction of those involved or add an unwelcome burden of responsibility depending on the employees involved.

Achievement

It is important that people go home at the end of the day feeling that their effort has been worthwhile. A clear understanding of the library's objectives and their purpose in helping it achieve them can assist.

Opportunities to learn and develop

Managers often underestimate the abilities of their staff and fail to offer them the chance to develop existing and new skills. Non-professional staff, for example, can be trained to answer simple information queries. their frequent handling of the stock can mean that they are in a better position to answer certain types of query than professional staff, although they need to be trained to recognize when to refer a query. It should be noted, however, that automation of some functions reduces the time operators interact with stock and thus the knowledge which was previously built up of both the stock and reading habits may be eroded.

Role under/overload

Everyone has an optimal rate of working, and it is desirable that people should be allowed to regulate their own work. As a rule, people achieve more, and work to a higher standard, if they are allowed to choose their own pace. Damodaran and Ip (1986) have observed how technical specifications can have consequences for job design and work organization.

If the pace of work is too fast for the operator, he may feel that he is constantly underachieving, and emotions of poor self-esteem and frustration may be experienced. Alternatively, he may try to increase his own pace of work. It has been found that if the system response time is consistently less than half a second, the operator's keying rate tends to rise above normal, with the result that muscular tension, stress and fatigue also rise, causing the operator to feel 'washed-out' afterwards (National Electronics Council, 1983).

If the system response is too slow the operator may feel he is overachieving and become frustrated that he is not in control of the pace of the work. In one text retrieval system studied by the authors, each record was verified, indexed and updated to the database after input, causing a delay of up to one minute before the next record could be entered. This caused the operator to feel that his time was being frittered away in short but frequent periods of idleness.

Pearce (1986a) comments on the effects of the software interface. 'User-friendly' systems may become irritating when greater skills are acquired: the option to 'bounce' through menus becomes highly desirable. This gives the operator the impression that he is deciding the pace at which work should proceed.

Kahn and Cooper (1986) mention that work underload has been linked with ill-health. Underload may be observed with many automated issue systems. The rapid issue and discharge of items which reduced queues at the counter, can also create periods of enforced idleness between customers. Unlike the manual system which it replaced, there are no odd filing jobs which act as a stop-gap. These problems may be exacerbated by technical faults or systems failures.

Role overload can also create problems. As staff costs increase, so it becomes necessary to think of automating the system rather than increasing the number of employees in order to maintain the existing level of library service. Undoubtedly the computer takes much of the workload, but can bring with it optimistic assessments of increased productivity. It also brings its own workload, needing to be fed data, maintained and backed up. Initially, too, extra work is required to train staff in the use of the system. In consequence, some people find that their workload has increased rather than decreased.

Role conflict

Where there are doubts as to the priorities in a job there is a risk of stress. The implementation of an automated system often creates more work for those in charge. Conflicts can arise between, for example, the need to create a usable staff manual for the system and the pressures of the other tasks the job-holder continues to perform.

Role ambiguity

This, another cause of stress, arises when there is doubt about what support will be available for decisions taken in the absence of a superior. Junior staff, for example, may be unsure how far to 'bend' the library rules if at all, or whether inflexibility will engender poor public relations. Lovecy (1984) comments that automated systems are more likely to pick up on infringements such as overborrowing which may make the library personnel seem stricter than before, creating its own conflicts.

The need for adequate variety

It is desirable for VDU operators to take short, frequent breaks away from the VDU. This can equally well be a change of task as a rest break. The reasons for this are threefold. First, an operator performing a visually demanding task will rapidly tire (see pages 47–48) and time spent working away from a VDU will enable the eyes to refocus and will rest the eye muscles. A 15-minute alternate task is recommended every two hours for operators performing jobs with moderate visual demands and every hour for people working on highly visually demanding tasks.

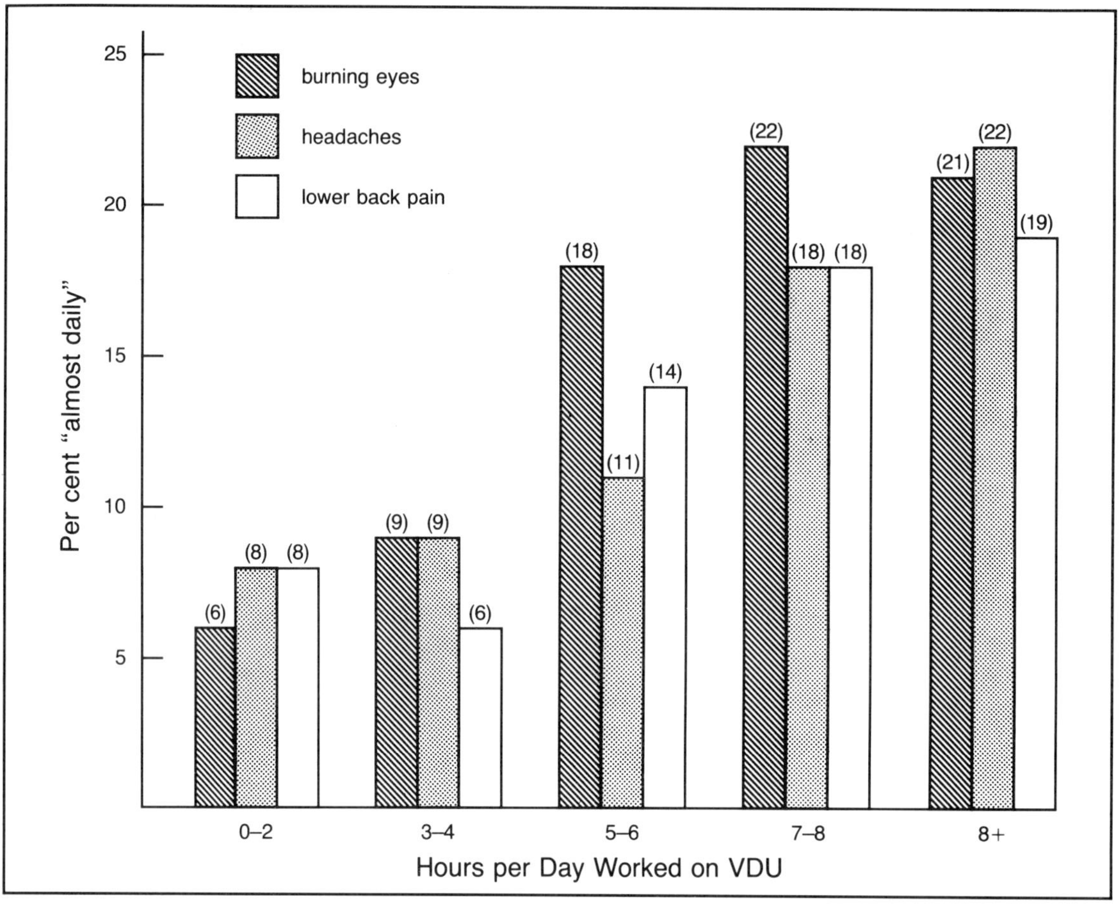

Figure 16.1 Percentage of workers reporting selected health symptoms "almost daily" by hours per day worked on VDUs
Source: Canadian Labour Congress as reproduced by the London Hazards Centre, 1984

Not only the eyes will tire during concentrated VDU work. Chapter 3 outlines the postural problems which may result from prolonged work spells (among other causes). Figure 16.1 indicates graphically the percentage of employees (the number studied is not given) reporting selected health symptoms. It is noticeable that as the number of hours worked on a VDU increases so, too, does the percentage of operators reporting problems, with a noticeable increase when more than four hours per day are worked. This implies that the total time spent on a VDU will affect the comfort of the operator and it is suggested that a total of no more than four hours per day should be spent on intensive VDU work. The graph in Figure 16.2 shows an increase in reported health symptoms as the length of time worked on a VDU without a rest break is extended. The implication from the graph is that if a rest break is given after approximately one to one and a half hours of intensive VDU work, fewer operators will report health problems. These graphs alone are not conclusive, however, as no details are available as to the workstation layout and the type of task being undertaken. Nevertheless, sufficient evidence is available (see Chapter 3) to show that alternate task breaks reduce the incidence of postural discomfort. Mentally, too, operators will tire if they are forced to work on one task for a long period. Attention will depend on an operator's motivation. If attention waivers, operator performance will diminish.

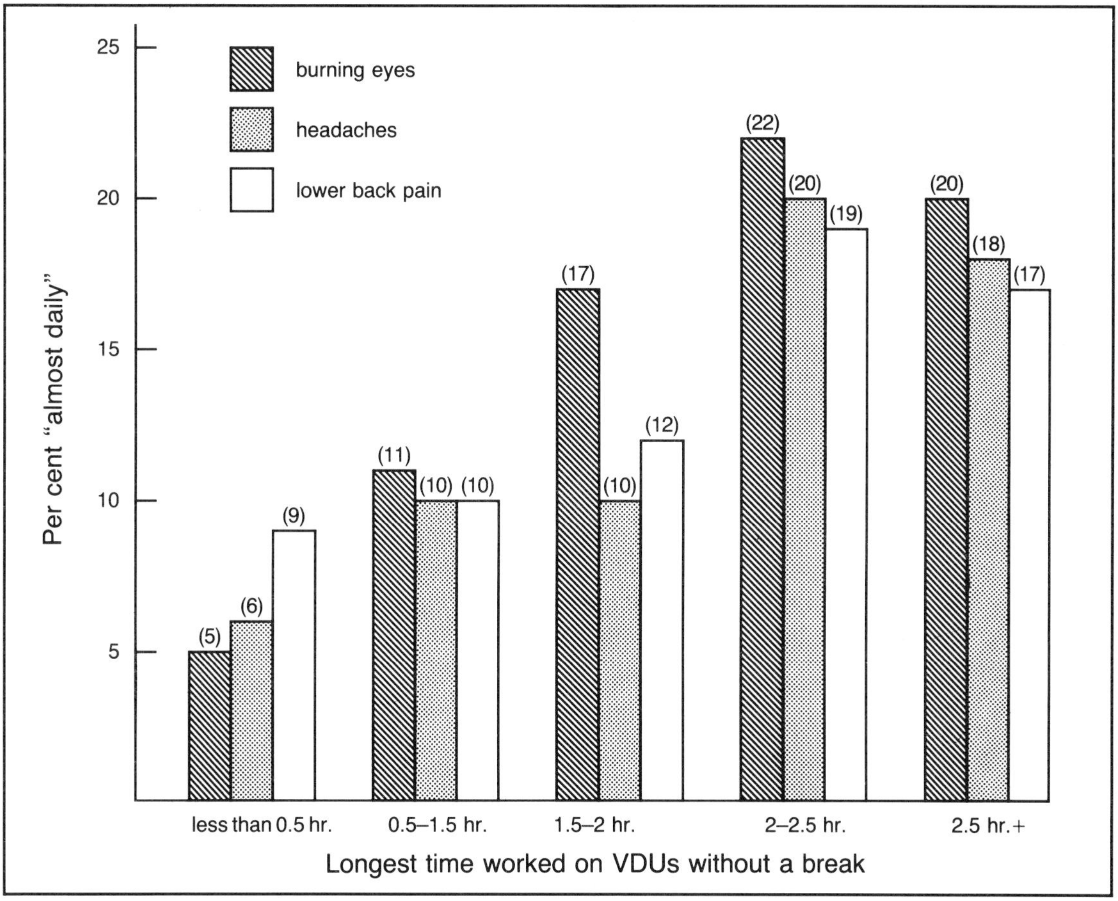

Figure 16.2 Percentage of workers reporting selected health symptoms "almost daily" by longest time worked on VDUs without a break

Source: Canadian Labour Congress as reproduced by the London Hazards Centre, 1984

Adequate changes in activity, therefore, are important for all VDU operators, but the length and type of interlude needed will depend on the operator and the type of task performed. As a general rule, work which is physically, mentally or visually demanding requires more frequent rest periods than less intensive work.

Techniques for improving job satisfaction

There are a number of techniques which can be used to minimize dissatisfaction and improve the quality of working life where tasks are intrinsically dull. These have been described by Argyle (1972) as job enlargement, job rotation and job enrichment, and are most applicable where staff have least control over their own assignments (although it must be acknowledged that, in the broader sphere, team staffing structures in public libraries are also a form of 'enrichment').

Job enlargement

Employees' jobs can be made to contain more tasks, and hence be enlarged. For example, the serials check-in clerk may be given, in addition, the task of assembling items for binding or the interlibrary loans clerk might be asked to collate statistics about the service.

Job rotation

Staff can be moved regularly from one job to another, bringing with it the advantage of increased job satisfaction. Job rotation has the added bonus of giving the employee experience of several aspects of library work and an overview of the whole system. This is particularly important for library trainees. Critics have rejected it on the grounds that each task is likely to be as boring as the last, and it is also difficult to determine the frequency of rotation. If a member of staff is transferred from section to section or department to department too often he may experience a sense of disorientation or that he has only just mastered one set of routines when he is once again confronted with the unfamiliar. If not monitored more closely, job rotation can therefore lead to dissatisfaction and additional stress. Nevertheless, well implemented job rotation can add variety and increased social contact to an otherwise routine job.

Job enrichment

Tasks can be organized so that the employee has some degree of responsibility. He may become a specialist in inter-library loans or the short loan collection and may acquire a certain amount of authority. Organizing staff into small teams in which they work together towards set targets can also give meaning and direction to their labour.

In conclusion, it must be said that uninteresting work may be unavoidable for practical reasons. Holroyd (1986) therefore recommends the techniques of job enrichment and semi-autonomous work groups. These are superior to the strategies of rotation and enlargement, he declares, because they involve an analytical approach to employees' work. The others are, in some measure, an admission of failure to create sufficient variety from within the job itself. An element of challenge is also a recognized criterion of good job design. Building this feature into basic library work which involves automated routines is very difficult. One way, however, is to emphasize the importance of the employee's work role to the overall organization. Counter assistants are ambassadors for the whole library service and should be acknowledged as such. Library assistant posts do not offer many chances for progression and personal growth. Management consequently has an even greater responsibility to ensure that other 'satisfiers' are met. There must be room for human interaction within the framework of the job. This can partly compensate for boring or repetitive work and is especially vital in a machine-dependent environment. Realistically, not all jobs will ever fulfil all principles of good job design, but, with thought, the quality of working life can be improved.

Automation provides an opportunity to improve the quality of working life. It is a chance to capitalize on what people find interesting and satisfying at work and to review total organizational efficiency at the same time. Sell (1986) sees it as a unique event because, with the introduction of computerization, employees' 'psychological contracts' with their jobs will inevitably have been broken and they will therefore be more receptive to change. He recognizes just how crucial is the management of change in this instance. To descend on staff and suggest that they have 'bad' jobs which need altering will not only be a blow to their own self-respect but will undoubtedly damage personal relations, perhaps irrevocably.

Any change must therefore be presented to staff in a positive manner. A consideration of job design is necessary in order to maintain morale. Unfortunately, this again is likely to be overlooked when time is short, and in the long term such an attitude can prove to be very expensive. Since involvement is crucial to success in this phase of the implementation process, the job redesign exercise should have a joint commitment from both management and unions. Damodaran *et al.* (1980) note that workers' participation in the redesign of their jobs is vital too, for they often place different value on aspects of their tasks which might not seem important from a cursory examination of their duties. 'Routine clerical procedures' and 'helping the public' are descriptions of the same thing, manager and job-holder interpreting the situation differently according to their emotional involvement. When the employee is actively engaged in work role analysis, the factors that motivate him will emerge: his new role can be designed accordingly to enhance his view of himself. Increased efficiency is not, in itself, a sufficient motivator.

Acknowledging that job redesign is a necessary complement to technology is not enough. Automation provides the opportunity for an innovatory approach but must also build on the past. This is particularly true where work groups are concerned. One of the reasons that people go to work is for social contact. Great dissatisfaction can result if the composition of any group is disrupted through change. Productivity levels may even drop. Finer (1982) stresses that existing relationships should be respected wherever possible. Nor is it advisable to reorganize tasks so that staff who dislike each other should have to work closely together. Wherever feasible, Finer continues, physical alterations to the library caused by computerization should take into account established 'territorial rights'. Compromises may be necessary but should be reached after amicable discussion.

It has already been intimated that the jobs of professional librarians are not (as a general rule) so adversely affected by automation as that of their junior colleagues. This is largely because the global function of their jobs incorporate more varied tasks and they can also exercise greater control over when and how to do things. The hypothesis is supported by a case study by Waters (1986) of the effects of new technology on employees at the University of Tasmania. A positive picture emerges from his (self-admittedly) limited survey: the minority of dissenting opinions coming from the non-professional echelon. Where problems have occurred at senior level it is usually the result of deskilling, and job redesign is an obligatory solution.

Management frequently has a carefully developed scheme for the production of advance new technology systems but no similarly imaginative plan for the design of future jobs and work organization. Corbin (1985) outlines a methodical approach:

- Prepare new job descriptions for each staff member whose duties and responsibilities will change once the automated system is installed.
- Discuss the new duties and responsibilities with the staff, stressing their continued importance to the library and the system.
- Orient each staff member in general to automation and the automated system.
- Train each staff member for his or her specific new duties and responsibilities.

Part VIII PLANNING AND IMPLEMENTING AN AUTOMATED SYSTEM

17 THE MANAGEMENT OF TECHNOLOGICAL CHANGE

The introduction of any automated system implies change. This may not be welcomed by the staff and it is necessary to understand the reasons why this might be. Managing the change requires good communications at all levels, the encouragement of staff participation, as well as selecting and implementing an appropriate system and training staff and readers in its use. The change to work patterns must also be recognized and good job design should ensure creative, satisfying work for employees.

Reactions to technological change

There has been an unprecedented escalation of technological developments in the last few decades, and it seems unlikely that this trend will alter in the foreseeable future: change, therefore, is a very real force in society. Burton (1987, p. 211) comments on the overall effects of that change: 'It is now almost a truism that the introduction of IT-based systems generates more conflict and stress than any other type of change'. This observation seems particularly appropriate for libraries where new and upgraded computer systems are constantly appearing on the market. In such an environment it is very easy to become obsessed and overinvolved in the technical aspects of information technology. To maintain a balance between human considerations and the technology of library automation, however, it is necessary to have an understanding of the complex and varied attitudes that people have towards change, and technological change in particular.

Fine (1982) comments that there are few changes in our social lives (other than war or other calamities) which seem to produce the same intensity of reaction that the coming of technology has aroused. The human response to technology may be one of exuberance, resistance or conflict. Many people have ambivalent feelings: computers are marvellous but, equally, disturbing. Turner (1981) makes a similar point. He acknowledges the infinite variety in staff attitudes to automation but goes on to suggest that the most remarkable feature about computers is their ability to provoke an emotional response from human beings. It is an ironic paradox that 'these pieces of mechanical apparatus which are designed to promote man's logical thought processes inspire the most irrational and impulsive behaviour from their potential users' (Turner, 1981, p. 13). This 'latent Luddite' tendency should not be dismissed out of hand because fear experienced by staff can create an aggressive or at best unhelpful attitude.

Any alteration in circumstances is almost invariably accompanied by an element of stress – this is true even when people have a positive attitude towards the change. The spectrum of responses will vary from mild apprehension mingled with excitement (which, in itself, can be stimulating) to fear.

Adverse reactions

People will show their opposition to change in many different ways. Some may complain loudly or predict failure whilst others attempt to ignore the topic. Reaction against technological change may be for one or more reasons.

Fear of the unknown

Uncertainty or fear of the unknown is a strong influence against change. Change is threatening to most individuals because no matter how unattractive the present may be, they have largely adapted to it. White (1985) notes that most individuals not only bend to meet the requirements of their job but that to some extent they also bend the requirements of their job to meet their own preferences – with or without their supervisor's knowledge and agreement. The perceived virtue of any long-standing system thus lies in its familiarity.

Technology represents change: it means giving up familiar ways and facing the unknown. The introduction of computer-based systems can thus be a stress-producing experience for the individual employee within the organization. Olsgaard (1985) attributes this stress (which can be either temporary or ongoing) to the sudden alienation of the employee from his environment. Even the most experienced employee may approach change in the form of automation with some trepidation because automation represents change in the way things have been done. White (1985) says that, quite aside from what automation promises to do for the library, individuals will be concerned about what it means for them.

Resistance to change is not exclusive to junior or middle grade staff. Rooks (1988) observes that change is often resisted by administrators, those who have the authority to accept or reject any proposal, which she calls 'resistance with veto power'. Again, the resistance may stem from uncertainty or fear of the unknown. The argument runs that, while there are acknowledged faults and inefficiencies in the present system, they are at least a known quantity. An overworked supervisor may only be able to envisage the problems of any new scheme and fail to see beyond them to the subsequent advantages.

Change is insistent and demands effort; it requires the acceptance of new ideas and concepts and it disrupts well established behavioural patterns, habits and securities. Stueart (1984) points out that the most vehement opponents of change are, not surprisingly, quite often those who have been in their particular job the longest. They have the most invested in the status quo and have fixed ideas on what is necessary, whether or not this coincides with any broad aims of the organization.

Fear of obsolescence and failure

Fine (1982) has stated that change inevitably results in a short-term reduction in our efficiency, calling self-esteem into question. There is a feeling that the knowledge and judgement acquired gradually over a number of years may become obsolete with the introduction of new systems and procedures. Experience suddenly seems to count for very little. Hand in hand with this fear of obsolescence goes the fear of failure, of an inability to adjust and do the new work. Status, privilege and authority are all threatened by lack of proficiency causing enormous apprehension and potential resistance to change.

With the introduction of computers, many employees may feel threatened with loss of esteem, position or power within the library. Hard-earned skills apparently lose their value. An example of this is cited by Clayton (1987): a traditional reference librarian may consider that his own professionalism is being undermined by the presence of a relatively junior member of staff with experience of online searching. Rooks (1988) articulates fears that the redistribution of information as a result of computerization can lead to loss of ownership. Staff who no longer find themselves in charge of critical information may experience a sense of loss or a diminishing of self-worth or value to the organization.

Bichteler (1986) recognizes the particular problems faced by managers who may have felt pressurized to purchase and use computers and who at the same time were suffering from similar anxieties. They may fear that many of their decision-making and supervisory functions may be taken over by sophisticated software, causing them to feel less important to the organization. One of the points raised by Mann (1984), in his survey of public libraries, is that many middle and senior managers received their training before the advent of computers in libraries and their lack of awareness may make them feel vulnerable to replacement by younger librarians who are more knowledgeable about computers.

Librarians may feel that their own importance diminishes in comparison to that of the system designers and programmers. Experienced staff may be largely ignorant in this particular sphere and consequently feel downgraded. Zoltan and Chapanis (1982) call this a fear of 'technocratic élitism'. Fear of obsolescence is perhaps greatest in middle-aged workers or those near retirement. They may feel either that retraining is not worthwhile or that management holds that view.

Uncertainty about the kinds of jobs created by automation can also be an area for concern. Staff may wonder whether the jobs will be as interesting and carry as much responsibility as before the change, or if the functions of one department will be merged with those of another. Lovecy (1984) comments that some staff will worry about the possibility of an increased workload. Many will be more concerned about their ability to master the new skills necessary for the job. Competence to operate the new system is, in fact, the principle fear identified by Sykes (1986) in his study of automation and non-professional staff at the Polytechnic of the South Bank.

Fear of economic insecurity

Employees may feel that automation is a step towards eliminating their jobs, if not immediately, at some point in the future. Economic insecurity is a prime reason for responding in a hostile fashion to new concepts. This fear is sometimes coupled with that of failure, the employee worrying that an inability to perform a task may lead to redundancy.

Automation has undeniably brought redundancies in some industries and therefore anxieties about job losses will exist. The extent of this fear will largely depend on the economic climate of the individual organisation and Lovecy (1984) admits that in academic libraries, where senior staff have contractual job security, it will mainly concern clerical grades. Redeployment, too, carries its own fears since the individual often has to adapt to a new set of colleagues as well as tasks.

Fear of changed relationships

The importance of social relationships in the workplace as well as in our private lives has been established by Argyle (1972). People may resist change because it seems likely that it will alter long-established work groups in some way. The individual member of a closely knit group may be carried along by the feelings of his peers in their attitude towards change. He, himself, may not be antagonistic but will not wish to hold out against an apparent consensus of opinion.

A library research project was conducted by Fine (1982) under the sponsorship of the US Department of Education to look at why people respond to technology in such complex ways. Library administrators, practising public librarians, library school staff and postgraduate students drawn from a national sample were interviewed. The conclusion was that the single most important fear that people have about technology is that interpersonal relationships will suffer: this overrode all other fears. Employees may be worried about who they will be working with, and who will be their supervisors, and whether their opportunities for interaction with other people will decrease.

Fear of computers

Many people experience fears related to the actual use of a computer. There is an aversion to physical interaction with the machine which Schraml (1981) connects partly to the workers' lack of understanding of the new technology and/or the decrease of workers' interaction with other people. She isolates three fundamental ways in which automation can seem dehumanizing: its pace; its intolerance of human error and the requirement that its operators communicate in a coded language. Ignorance of how computers work often leads to a feeling that one wrong move could bring the whole system down. Lovecy (1984) identifies several common concerns: lack of a printed record; unwittingly wiping out a set of data; corrupting data; and damaging equipment. Some people never learn to trust the machine and wish to check and doublecheck its operations at all stages. Sykes (1986) confirms that nearly all the non-professional staff interviewed had been apprehensive about the prospect of automation. While several library assistants talked of sleepless nights before the system went live, mention was also made of two older members of staff who left shortly after the computer arrived, largely because they could not overcome their fear of it. The fears centred around inadvertently harming the computer or computer records. An investigation by Lefebvre (1985) into training library assistants for automation reveals a similar picture. Concern voiced by the interviewees was that they might 'destroy all the records' that 'something disastrous might happen' and that they might 'ruin the program'.

Computers also carry their own mystique. The complicated equipment, jargon and very cost can all be off-putting to the individual lacking self-confidence. Mistaken impressions can exist about the level of expert knowledge required to operate computers.

Other specific fears that people have may relate to health factors. The main worries about operating visual display terminals (VDTs) as identified by Cross and Bawden (1987) are: radiation hazards; visual problems; musculoskeletal problems; tenosynovitis – repetitive strain injury; facial dermatitis; psychological and psychosomatic problems, all of which are discussed in detail in Part II.

Fear of increased management surveillance

There are also concerns that the introduction of computers may offer an opportunity for increased management supervision and surveillance. Errors and inefficiencies become more apparent in automated systems and output is more easily measured. This facility may be seen by some employees as a form of 'Big Brother' tyranny. Readers, too, may resent personal details being held on computer and worry about the use that is made of them.

Positive reactions

Conversely, the positive response to automation is not without its problems. Fine (1982) describes how an 'overboard positive' reaction can be as damaging as a negative one. Members of staff who throw all their energy and commitment into a new scheme and totally abandon the old also create difficulties for management. There is usually a lengthy time lapse between deciding on a replacement system and implementing it and during this period service has to be maintained.

Management of change

Rooks (1988) observes that change is mandatory if an organization is to survive in a world of developing technology. Responsiveness to the demands of an organization's clientele and its employees also implies change. In a volatile environment, skilful management is all important. Jones and Jordan (1987, p. 8) comment: 'Automation makes it necessary for librarians at all levels to be more conscious managers, if their staff and users are to accept changes positively.'

Many of the staff fears relating to regrading of jobs and deskilling of tasks are perfectly legitimate. Fine (1982) notes, however, that it is imperative to acknowledge that all human behaviour has value. Resistant responses are far from being totally negative. They serve a purpose in causing management to re-examine proposals and act as a natural check on too avid an acceptance of innovation for its own sake.

In any situation it is far easier to prevent fear and adverse reactions than it is to remove them once they have become established. The library manager should therefore anticipate responses to the introduction of new systems and wherever possible take appropriate action to reduce stress. Clayton (1987, p. 167) observes:

> Change should be seen as a way of improving existing services and of being able to provide new ones, that is, change as a provider of challenge and opportunity rather than as a threat.

Library automation is a complex and demanding management task. Turner (1981) emphasizes that there must be a mutual sense of purpose and a mutual sense of direction between all involved for the project to be a complete success. He gives four important pointers for creating a positive attitude among staff which can be equally applied to all users. They should become:

- confident
- informed
- aware
- trained

The methods which can be adopted to achieve these aims and to prevent or at least minimize anxieties towards technological change are explored in the following chapters.

18 THE IMPLEMENTATION PROCESS

The implementation process involves a planning stage and an implementation phase.

The planning stage

The success of a system is not dependent merely on whether it is technologically sound, but also on staff and user reaction. Planning for automation therefore involves planning for the people and their interaction with the computer. This means taking into account the human factor in the choice of system: the hardware; the software interface and any consequent furniture requirements, which have already been discussed. It also implies preparing staff and other interested parties for the changes that computerization will bring. It is therefore necessary to consider who needs to be kept informed of the project and its progress; when they need to be told; what they should be told (and why) and how the information should be imparted. Some idea of the timescale should also be available. Boss (1984) states that a single promotional technique is not enough. Different audiences will naturally have different information needs and each should be considered on its own merit. This stage in the implementation process is vital: according to Clayton (1987) it has not been unknown for staff to refuse to operate newly installed systems because they have not received union approval. Keeping all those with a vested interest in a project informed tends to generate goodwill in itself and valuable reciprocal advice can be received.

The individual information requirements of various groups are further explored in this section. It must be stressed that the order in which they are studied does not necessarily imply a hierarchy of need since much communication should take place simultaneously.

The funding body

One of the first tasks is to convince the funding body that the proposed automation is worthy of formal investigation. Justification is usually in terms of financial benefit and savings (presented in report form), although care should be taken to avoid over-enthusiastic estimates that may lead to subsequent staff or bookfund cuts. Possible extensions to the service should also be noted.

Clayton (1987) suggests that a good way to 'win friends' is to arrange a demonstration of equipment. Nevertheless, she warns, it must be well organized in order to serve its purpose; invariably something goes wrong when the occasion is important. The operator should appear confident and be able to meet all contingencies, otherwise the demonstration may have a negative effect. Demonstrators must convey their enthusiasm for the system but not forget that they are trying to convince users, rather than librarians, of its advantages. The talk should be pitched at an appropriate level. One method of personalizing demonstrations is to include names of participants in a borrower database: a simple but successful ploy even at senior level. Lovecy (1984) notes that the feature most appreciated of the SWALCAP circulation system, when demonstrated to members of the library committee at Reading University, was the message that an item had been returned at a specific time on the same day. This provided readily comprehensible evidence of the difference between an online, real-time system and the previous offline batch system. Administrators may also be impressed by ways in which the system can be used to serve management purposes through the generation of statistics. The advantages of tighter stock control can be emphasized.

Well thought-out and detailed communication at this level helps build support for the project over and above the purely financial. Influential friends and allies can be won who may be able to exercise their authority at critical stages later on. Once implementation is under way, management meetings and bulletins can be used to inform officials of the progress of the development and anticipated completion dates.

Staff

An efficient automated system will only come about if the staff who operate it are happy to work with it. Negative reactions to technological change have already been discussed but it is necessary to recognize that the underlying cause of all these reactions is fear. Uncertainties and fear are further compounded by rumours and speculations which spread in the absence of hard information. Thus it is argued that effective communication in the planning stage is one of the best ways of conquering misplaced fears. Quinlan (1987) observes that no organization should consider embarking on major change without first closely examining the effectiveness of its communication systems. Handy (1981, p. 353) warns that 'if there is one general law of communication it is that we never communicate as effectively as we think we do'.

The information passed on to staff should address expressed and unexpressed fears, most of which are possible to anticipate. For example, reassurance should be given about personal matters such as job security and job content. Complete honesty and openness are crucial. Reeves (1984) states that, in the automation of Essex County Libraries, any reductions in staffing levels were dealt with by natural wastage. This is probably the least distressing solution for all those involved, but exactly what is meant by the term should be carefully explained. Staff should not think that pressure will be exerted upon them to leave. In the exceptional event of redeployment or redundancy, Lovecy (1984) advocates involving the organization's personnel department.

Other specific fears should be anticipated and tackled – for example, health and safety, and the use of work monitoring. Lovecy (1984) recommends that the manager should admit if the latter facility exists and explain under what circumstances (if any) he would use it. Confirmation should be made that adequate training will be given.

System maintenance and failure can create a lot of stress and staff need to be warned that problems are likely to occur, especially in the early days. Often there can be problems getting through to someone who can help and, even then, there may be considerable delay before the fault is found and rectified. In one library it took a year to get the system running because of an incompatibility between the network and the system, neither supplier being willing to accept the responsibility.

As automation should bring benefits to the library, these should be outlined – especially how the new computerized system will help staff in their work. When the potential advantages are appreciated there is more commitment to the change. White (1985) maintains that such a perspective helps to prevent any feeling of lament for the old way of doing things. Staff are also in a better position to deal with readers' queries if they themselves are up-to-date with developments.

Although the positive aspects of automation should certainly outweigh the negative, staff expectations must not be raised unrealistically. Promises that computerization will make the tasks of library assistants more interesting and rewarding may prove to be rather hollow. Computerization is not a panacea for all ills and there are still boring jobs to be done. Atkinson (1984) advises that when we plan for change we have to accept a certain error rate. No change will be perfect and where computers are involved there will be a certain amount of downtime, a certain amount of failure. It is unlikely too that the full potential of any system will be realized in the first instance. Acknowledging early on that there may be unexpected setbacks helps prepare the ground for when they do occur. If unbalanced claims are made, then staff will be disappointed and may become disenchanted with the whole scheme.

A variety of communications methods are available:

- written communications, such as news bulletins and newsletters;
- lectures;
- visits;

- departmental meetings;
- an automation working party;
- participative management

Written communications

News bulletins are a popular way of disseminating information, especially when large numbers of staff are involved. They are useful in dealing with both general topics and the more specific. Herring and Mackenzie report that this was the principal method of communication used in Aberdeen City Libraries before automation, where separate leaflets dealt with barcode labels, overdues and reader registration. The language used in the bulletin should be fairly simple and straightforward with an explanation of any jargon, otherwise they can be self-defeating. Herring and Mackenzie (1986, p. 28) give an example:

> Book cards and seried ranks of issue trays will disappear, as will, in time, the card catalogue as we know it. In the future, every book will have a date label as now,plus a 'bar-code label' (you've probably seen them in supermarkets) on which a special code represents a 14 digit number. This will be the new book number and will supersede our present accession number.

Quinlan (1987) has remarked on the problem of achieving a happy medium between satisfying people's information needs and inundating them with so much paper that the exercise becomes meaningless. Written communications should constantly be reviewed and improved, and the opinion of the recipients sought on the usefulness of their content. Another drawback to this type of communication is that it is essentially passive. Recipients do not have the chance to react to the information in two-way discussion; consequently it should only be used as a support to other methods, not as a substitute.

Lectures

Lectures are also, for the most part, a passive form of communication but are useful for transferring information to several people simultaneously and for explaining complex issues. They are particularly useful for training.

Visits

Visits to other automated libraries provide staff with the opportunity to see a working system in operation. It is helpful if visiting staff can talk to their opposite numbers using the automated system to assess the impact it will have on their own jobs. Lovecy (1984) warns that such a visit should be carefully planned: the library selected should be one in which there is a positive attitude towards automation. Some sort of prior preparation should take place. Staff will get more from a visit if they know the type of things to ask and they can then report back to others who were unable to go. Visits to other places provide the opportunity to create a team spirit and sense of solidarity. Some people find that seeing the system in operation reduces their feeling of alienation towards machines.

Departmental meetings

Staff meetings are a useful method of communication but can also tend to be one-way unless the management is open to comments and suggestions.

Automation working party

It may be desirable to set up a working party comprising representatives of all levels and sections for the express purpose of planning the new automated system. Ideally, the members of the working party should reflect a spectrum of views from the enthusiast to the 'Luddite'. Baddeley (1988) suggests identifying individuals who can act as 'missionaries' to their peers. Different interests are therefore represented and benefit may be derived from the exchange of ideas within the group. Individuals or subcommittees can concentrate on particular aspects of the automation plan, such as the requirements of different functions, health and safety aspects, furniture and so on.

The project head should command the respect and loyalty of both staff and management. He should be able to maintain a broad overview of the development whilst exhibiting a detailed and sympathetic understanding of human concerns. In small libraries the 'project team' may only consist of the librarian and one other member of staff but a considered approach towards involving everyone is still necessary. A cooperative working relationship is highly desirable.

Participative management

Participative management is seen to involve a wider group than the automation working party. The subject has received much attention in the literature, and it has generally been found that job satisfaction is greater when people are allowed to contribute ideas (Marchant, 1982). Dakshinamurti's survey (1985) to study the reaction of library personnel towards automation found that the most desirable characteristic in a manager is a participatory management style. Decisions that are reached through participative rather than hierarchical management are of a higher quality and have greater group acceptability, particularly when the composition of the group is representative of different areas of expertise and points of view (Sidgreaves, 1985). West (1987) refers to the 'emotional investment' that people have in decisions to which they have contributed and this, in turn, encourages them to share the results of discussions with their colleagues.

Where true participative management occurs, decisions are left to those with the most relevant first-hand experience, a principle endorsed by James (1987) in his account of automation at Kingston Polytechnic. The participation of non-professional staff was actively sought on decisions relating to the system as a whole and particularly on those which affected a task which they performed; the rationale being that the junior staff would have greater knowledge of the individual clerical routines. James argues that such involvement is effective management rather than democracy.

Lucas (1974) has summarized the benefits of staff participation in the management process as follows:

- Participation can be ego-enhancing.
- Participation can be intrinsically satisfying and challenging.
- Participation results in more commitment to the change.
- Participants are more knowledgeable about the innovation.
- Participants can observe better solutions to change problems.
- Participation means that the user has retained some control over a system that will affect him.

The concept of participative management, however, does not have universal and unequivocal acceptance. Some hold the view that the involvement of all staff is unnecessary because the management team contains adequate expertise. Marchant (1982) acknowledges that staff participation consumes valuable working time. Often, too, funds may have to be spent within a restricted time period, rendering user involvement impractical within the time allowed.

It cannot always be assumed, either, that staff will want to participate. Sykes (1986) found that junior staff at the Polytechnic of the South Bank did not want to be included in the debate about the system's design. They had insufficient knowledge about automation to make constructive suggestions and, while their understanding could have been improved through visits and lectures, this was not thought to be economically viable. Where a second or third computer system is installed, however, Sykes concedes that junior staff could prove to be amongst the shrewdest critics of the existing system.

Staff involvement in the planning stage will therefore be affected by various factors, such as time available, the size of the library and the willingness of staff to become involved. While direct participation of all staff and other interested parties is not always feasible, it is important that some form of two-way discussion is established to keep staff informed about the progress of the project and to enable them to air their ideas, their hopes and their fears. Bichteler (1987) makes the following recommendations:

- Involve staff initially, telling them why, what and how automation should take place.
- The planning committee should include employees of all levels, especially people who are influential among clerical and paraprofessional staff and sceptics.
- Organize demonstrations and visits for staff.
- Proceed slowly, allowing time for feedback and let the staff determine as many options as possible.
- Work carefully with those whose office space and routines may be disrupted.
- Inform everyone completely and frequently of progress, decisions and specific plans for training.
- Promote the system. Tell staff how it will relieve them of chores.

- Prepare for a lot of hand-holding for nervous staff members when the system goes live.

Unions

The trade union movement has always sought to protect employees from the adverse effects of new technologies, library automation being no exception. The widespread use of computers in the early 1980s encouraged trade unions to look at the situation more closely. Guidelines for negotiating 'New Technology agreements' with management were established. Jones and Jordan (1987) set out a comprehensive list of points on which agreement between management and unions have to be reached before implementation of new computerised systems can go ahead:

- An assurance that jobs will not be lost.
- Agreement by management to retrain staff to use the new equipment.
- Divulgence by management of future plans for introducing new technology.
- Permission by management for trade union representatives to attend new technology courses so that they will be adequately informed on the implications of introducing new technology.
- Where redeployment or alternative work becomes necessary, existing staff are to be protected, and joint management/union discussions held on suitable alternative work.
- Where new technology gives certain staff new responsibilities and demands new skills, this should lead to discussions on regrading or additional pay.
- The level of job satisfaction should be safeguarded, and tendencies towards deskilling should be monitored.
- Negotiators should insist on certain standards of health and safety for staff working with VDUs, and monitor the effects on staff conditions in the workplace.

Data protection is also covered by some new technology agreements, enabling a check to be made on the records that organizations keep of their staff.

Union representatives should be involved from the beginning and should be free to decide their level of involvement. Failure to consult can result in setbacks. Dean (1987), County Librarian of Durham, reports a delay of almost two years while agreement was reached with trade unions on the expansion of their automation programme. Unnecessary bad feelings can be generated when consultation does not take place: unions can even prevent systems from being operated. Early involvement of staff and unions is the key but, as Clayton (1987) points out, much will depend on the prevailing local atmosphere in the field of industrial relations.

Consultation with the unions can bring with it the benefit of the considerable help and advice some unions are able to offer, especially in the areas of health and safety and job design. Many unions issue leaflets or information packs which can be given as additional information to staff, but it may be advisable to check for any bias and to supplement the leaflet with additional information if need be.

Library users

A central argument for introducing computerized systems into a library is the improvement that it will bring to reader services. The readers should not be left to discover these benefits for themselves, rather they should be informed through a well constructed publicity campaign, which in turn will help to increase their acceptance of the system.

Library users will need to be told why automation is necessary, how it will affect services and when it will take place. The benefits, such as the savings in time and improved access to the stock, should be spelled out. They will also need to be informed of any requirements the system will make of them, such as the need to bring their ticket with them in order to borrow books.

Whilst the benefits need to be outlined, the dangers of oversell must be avoided. An honest, but positive, tone is recommended. Lovecy (1984), for example, argues that there is nothing wrong with boasting that a microfiche catalogue will provide an up-to-date record of stock, providing that the drawbacks of splitting the catalogue (where it is necessary to do so) are likewise mentioned.

Too little information can also be dangerous, a point expressed forcefully by Lovecy (1984). He argues that, when readers are told that the microfiche is merely an interim measure before the introduction of an OPAC but given no idea of timescale, they may make aggravating demands for the *proper* catalogue. This in turn can be demoralizing for the counter staff who bear the brunt of the comments. The possibility of downtime should also be mentioned, thus preparing readers for the shaky start that so often occurs with a new system. Assurances should be given that there are adequate back-up procedures in this eventuality.

One of the simplest and most effective methods of presenting this core information is to produce a library leaflet outlining the automation plans, such as that issued by Oak Lawn Public Library to introduce their new issue system (see Figure 18.1).

Alternatively, any newsletter which is regularly produced should include a section on automation, such as those issued by Hillingdon libraries. Leaflets need to be attractive and easily understood so care should be taken to avoid the use of technical wording and abbreviations (many readers, for example, may not know the meaning of 'VDU').

Leaflets can be supplemented by displays, posters, bookmarks and audio-visual presentations. The public library may take the opportunity of any coverage in the local press to remind people of its existing services. Progress reports are essential in any case but particularly so when there is direct inconvenience to the reader. Special graphic charts may be used to illustrate how much of the bookstock conversion has been completed and to give an indication of how much longer disruption can be expected.

Figure 18.1 Promotional leaflet designed to encourage user acceptance of a new automated issue system. Courtesy of Oak Lawn, Public Library, Illinois, USA

Readers need to be advised well in advance of the system going live, especially if they will need to re-register or be issued with a new library card.

Installation

Planning for setbacks

Thorough planning and involvement of all those affected by automation should go a long way towards making the actual installation of the system relatively problem-free. However, it is almost inevitable that unforeseen setbacks will happen. The most organized library has to rely on outside agencies for the delivery of equipment, and delays in its arrival frequently occur. Problems are further compounded when the equipment is found to be faulty or incomplete. Testing of hardware and software can be a difficult time for the library and is often a lengthy business. This is the point at which even the best timetables tend to go awry.

Procedures must exist for informing staff and users of the reasons for delay, especially if it inconveniences them in any way. An account of what is being done to remedy the situation should also be given and a revised timescale suggested. Although information of this kind cannot dispel doubts about the future operational reliability of the system, it can at least help to keep the issue alive. Damage to public relations can be minimized by broadcasting the facts.

Planning for disruption

Any computer development is bound to cause a number of disruptions to the normal running of the library. These include discussions with library staff with respect to the system specification, the physical changes which need to be made to the workplace, the installation of new furniture and equipment, the staff training programme, and the testing of the system (Damodaran *et al.*, 1980).

Staff need to be aware of the possibility of disruption before it occurs and patience needs to be exercised all round. Productivity is likely to decrease over this period and due allowance needs to be made. A number of practical points need to be considered.

- Time needs to be allowed for staff to adapt to the new working environment and to mourn for the old (Nadler, 1980, p. 6):

  > People need to mourn for the old system or familiar way of doing things. This is frequently manifested in the emergence of stories or myths about the 'good old days' even when these days weren't so good. The process of dealing with loss takes time, and those managing change should take this into account!

- Adequate alternative workplaces need to be made available when physical alterations are made to enable staff to continue their work with minimal inconvenience.

- Health and safety factors should be considered. Boxes should not be left lying around where they are likely to constitute a hazard, either through blocking passageways and exits or protruding where people are likely to trip over them. The same care should be taken of discarded furniture.

Installation can be a hectic and fraught period for all concerned. Fine (1986) stresses that people's dependency needs are greatly intensified during change. They seek support, feedback and direction from one another and from those in charge. Communication, which takes on an even more vital emphasis at this time, is less easy to achieve. Gratton (1983, p. 23) makes a perceptive remark on the situation in Derbyshire County Libraries:

> From the outset it had been the intention to keep staff fully informed as ideas and activities progressed. I have however to acknowledge a degree of failure . . . [because] 'At the time of greatest need for communication there is the greatest shortage of time to communicate.' When things are happening everyone appears, not surprisingly, to be wholly involved in making them happen.

Changeover procedure

There are several different approaches to the changeover from one system to another which can be employed: some cause less problems for staff than others. Tedd (1980) identifies four methods: complete changeover; phased approach; parallel running; and pilot operation.

Complete changeover

On a predetermined date, one system is completely replaced by another. This has two main advantages. It is cheap (the library does not have to meet the cost of two systems at the same time), it can be quickly effected, and the benefits can be rapidly appreciated. On the negative side, it is only really practicable if the system is simple and not too many departments are involved. Also, it offers no 'safety net' if anything goes wrong, and it does not give employees time to adjust. A complete changeover must be supported by excellent back-up facilities.

Phased approach

The system is divided into sections and each is installed as a separate unit. This enables problems with one module to be resolved before another is introduced. A controlled, gradual method, the phased approach allows operators time to adapt and to learn, and also reduces the likelihood of a complete system's failure. Nevertheless it can make the implementation process rather a protracted one and the full benefits of the system cannot be felt until the changeover is complete.

Parallel running

The old and new system run side by side until the new one is performing satisfactorily. The advantages of parallel running are that the library service is uninterrupted and, in an emergency, it is possible to revert to the old system. However, it is an expensive solution and staff may become confused and frustrated over how to use each system. Lack of space, for example, may also inhibit the simultaneous operation of a Browne issue system and VDU input.

Pilot operation

The system is tried out in one small branch before being installed in the main library. Difficulties can be sorted out with less embarrassment in a 'guinea pig' branch and the prestige value of being the first to acquire a system may also offset some of the accompanying traumas. Public relations, too, are easier to safeguard on a smaller scale. This is a useful approach for a multi-site environment.

Several factors will determine which changeover method is preferred, including the particular characteristics of the system to be installed, the type of library, and any time constraints.

Evaluation

Whichever changeover procedure is adopted there will be some relief when the system is finally up and running, but it must be recognized that this is not an end in itself. Evaluation mechanisms must be devised to monitor progress. These can be formal or informal, depending on the situation.

It is common practice to follow installation with a trial period during which 'bugs' in the system can be removed and various adjustments made. Difficulties in using the system or errors should be reported and logged. This is an essential part of the overall plan. Quinlan (1987, p. 36) comments that 'The trial period should not be shortened to offset delays in earlier parts of the project'. Above all, the process of communication should continue: the systems librarian will want to make adjustments to the program and users will have their own ideas to contribute, especially as their confidence in using the system grows.

No system will remain static. The person in charge of the computer will have information to communicate to staff, not just on the faults of the system, but on new ways of using it. Damodaran *et al.* (1980) comment that details about major changes should be passed to users through a formal training programme. Smaller changes, they suggest, do not warrant the same approach, although users should know of the change and its implications before it takes place and they should have the opportunity to ask questions. Accompanying documentation should also be available. The need for setting up a structured or informal vehicle for feedback from the users is often overlooked, problems being dealt with locally as they occur. However, it is desirable that all staff should be able to learn from the discoveries made by their colleagues.

Comments from all staff should be given careful consideration, not just those from staff who use the system regularly or organize its use by others. Lovecy (1984) points out that the reaction of an employee unfamiliar with the system, or a new staff member, may give some indication of how an untrained library user might perceive the system.

Evaluation needs to be carried out with sensitivity and tact. On an informal level it can be done by watching operators at work. Clayton (1987) admonishes that techniques for measuring systems' performance may hint at 'Big Brother' tactics. When the time taken to perform tasks and productivity are under scrutiny, staff must be aware of the reasons for it or they will misconstrue management's aims. If tasks are taking an inordinate amount of time this is probably due to as yet unidentified faults in the system. The mutual benefit to all in isolating these faults should be stressed. The best way of avoiding suspicion is to involve the staff in the evaluation exercise. Sensible suggestions must be followed up whenever possible.

A less tangible, though equally important, gauge of the system's performance is employees' morale, which in turn reflects their job satisfaction. Any sudden, unexplained increase in absenteeism or poor timekeeping when computerization is introduced may need further investigation.

The reaction of the library's clientele is also significant. Complaints and praise alike can be valuable indicators, although here there is an even greater communication problem – unless people feel very strongly they are unlikely to voice their opinion effectively. Comments books are one way of establishing views, although they are frequently open to abuse. Marchant (1982) states that, as a library gets larger, the librarian's direct contact with its users decreases and he must therefore rely on his own staff for reader feedback. This assertion brings the discussion full circle: staff are unlikely to relay readers' comments unless they are already confident that their own remarks are paid serious attention.

Time must be made to appraise the project's success or otherwise. It is perhaps best to do this after a year or so, when much of the heat should have disappeared from the situation and constructive comments can be made. Future plans should also be outlined at this time. At this stage, a more formal method of evaluation may be required. Interviewing the staff and users and holding formal meetings to discuss the system are two methods which may be used. The nature and the size of the library will obviously have an impact on the most appropriate method chosen.

19 TRAINING

Education and training can be thought of as an extension to the communication process: they are a means of informing people in more detail of how changes will affect them and of teaching them to cope with these changes. They are, in turn, a logical way of overcoming fears and misconceptions. Thorough training is of great importance. To a large extent the effectiveness of a system will depend on those who use it, and adequate training is needed in order to exploit the system's capabilities. Well trained staff are generally confident in handling the computer, a confidence which transmits itself to the users. Conversely, poor or insufficient instruction creates bad relations between people and the computer which can lead to feelings of stress. Automated library systems are very expensive, so too is quality training, but there seems little point in spending large sums on the one without a commensurable outlay on the other.

The staff training programme

The training needs of staff must be carefully analysed well before the installation stage. Again, managers must ensure that their organization does not merely pay lip-service to the idea. A comment by the Library and Information Technology Centre (1987, p. 86) is worth bearing in mind:

> The scale and success of training appears to vary enormously and there is often a considerable gap between the stated aims and objectives of particular programmers and the actual implementation.

Dakshinamurti (1985) found that library personnel placed much emphasis on training. The staff training programme will vary according to time and funds available, the abilities of staff, the scale of computerization, and the size of the library service. Some of the common features which can be identified are analysed below.

Trainees

The manager's first task is to identify all the staff who will need some form of training. There appears to be a general consensus in the literature that this should have as broad a definition as possible. In addition to everyday users, trainees should include those who may use the system occasionally as well as those who work in the library but do not have to operate the computer. However, it should be noted that the same type of training will not be appropriate for all. If staff are not involved in the training sessions, they may feel ostracized, which can have a detrimental effect on moral. Bichteler (1986, p. 126) expresses it thus:

> The [people] who don't use the computer feel socially isolated since they can't contribute their 'war stories' . . . Psychologically, it would be valuable to train everyone.

Problems may also occur when staff change jobs within the library system, and with readers. Readers seldom discriminate between library personnel, and anyone working in the library is likely to be asked for help in using an OPAC.

Particular attention should be paid to training senior and middle-grade personnel, as they may have missed out on opportunities for IT training and education earlier in their lives. Training senior staff first makes managerial sense, since they are then in a position to give direction and support to the project. Such a policy will help to prevent the feelings of inferiority that may arise if all grades are trained together. The sense of inferiority may be on either side, since junior staff often learn more quickly than their older colleagues: damaged pride should be avoided.

Where time is short, however, top priority must be given to those who have to use the computer in their daily work. The needs of part-time and evening workers should not be overlooked, and special contingencies should be made to ensure that employees do not miss out altogether on training through illness or leave.

Trainers

The training provided by systems' suppliers varies in extent and quality. Dyson (1986) emphasizes that this must be thoroughly investigated when negotiating the contract. She asserts that the 'when and where' of training should be decided by the individual library, and not by the commercial organization but, in reality, a compromise may have to be sought. System operator training is usually limited to a few staff, but it is important that, even in a small library, a couple of people receive this type of guidance to ensure continuity if one person leaves or is away for any reason. Some contractual arrangements with the vendor provide for general training of staff on site. This may have its own drawbacks. Sykes (1986) reports that the GEAC demonstrator who visited the Polytechnic of the South Bank seemed unfamiliar with library procedures .and practices. Use of jargon is also very frustrating at the learning stage. Instructions such as 'move the cursor to the subcommand field' and 'wand the barcode' are not very helpful (Sykes, 1986).

Staff may feel happier learning from people whom they know and who will have some idea of their abilities. If the trainers are themselves unsure and hesitant, however, they will only perpetuate unease with the system. Trainers must be competent at their task and able to communicate well. When training is provided in-house, managers must allow ample time for trainers to familiarize themselves with procedures, otherwise it will be a false economy. Lefebvre (1985) remarks that the value of advice to be gained from other libraries which have automated using similar systems should not be forgotten.

Location

Where the training takes place will largely depend on who is responsible for training. There can be logistical difficulties involved in travelling to a firm's headquarters for a demonstration. Corbin (1985) advocates that training should take place in the library and on its own equipment wherever possible. There is a sense of security to be gained from the known environment. The practicality of this will very much depend on the premises available. Above all, staff should be removed from their normal duties in order to give their undivided attention to the training.

Timescale

Different aspects of training can take place at different times. General orientation cannot be started too early. According to Damodaran *et al.* (1980), more specific training should occur just before the skills it is imparting are to be used. The logic of this is frequently defied in practice. Training that is left too late can be rushed (if not foregone) in the implementation process. Examples abound. Bagley and Oyston (1982, p. 51) remark on the situation at Hatfield Polytechnic:

> Plans for a comprehensive training programme for library staff were seriously affected by the unavailability of the system until several days before the beginning of the new academic year.

Staff may be apprehensive enough at this juncture without worrying whether they are actually to receive any instruction.

It is difficult and unwise to predict the exact time required to train individuals. Dyson (1986) says that 30–36 hours proved sufficient at Westminster (which included 18 hours of practice and revision), but there are too many variables to give even a rough estimate. What must be stressed is that sufficient time should be allowed and physical and mental overload avoided. Learning anything new is exhausting. Concentration can only be maintained for short periods. Lovecy (1984, p. 168) comments: 'People take quite a while to familiarise themselves with a system, and even longer to convince themselves that they know it.' This is another reason why training should not be a last-minute affair. Sufficient time should also be left to change training methods should they prove inadequate or unsuccessful. Burton and Petrie (1986) maintain (perhaps rather idealistically) that the changeover date should be amended if training has not been completed on schedule.

Content

The same level of training and education is not appropriate for all staff within the organization, although everyone should be given some appreciation of what computerization can do for the library. Determining just how much background information to give staff on how computers work is a contentious issue. On the one hand, it is argued that such knowledge will help to increase their understanding of why certain things happen (or fail to happen), while on the other it is seen as an unnecessary burden likely to cause even more stress. Fine (1982) observes the paradox in the move towards user-friendly terminals. They diminish the need to have any real grasp of how the program functions but in doing so strengthen the belief that automation is really beyond the comprehension of most of us.

Sykes (1986) expresses the opinion that sending junior staff on computer courses may be intimidating and counterproductive. Reassurance should be given that no knowledge of computing or mathematics is necessary in order to do their job: technical lectures may give them the opposite impression. Other authors, notably James (1987) and Lovecy (1984), advocate increasing knowledge on the basis that staff are then in a better position to form intelligent answers to readers' queries concerning the computer. Lovecy (1984, p. 172) stresses the difference in stating 'It says CPU unavailable – I don't know quite what that means' and saying 'I'm afraid that they're having problems at the central site: we shall either be able to regain control in a minute or we shall start recording issues locally.' The second reply is far more positive. It helps to promote a good library image and reduces the operator's own sense of inadequacy.

Some understanding of what computers can and cannot do prevents the situation where staff blame faults on the computer. Computer 'errors' are far more likely to be human errors. A fear identified in Chapter 17 was that of data corruption. An explanation to employees of inbuilt checks in the system itself can remove these unfounded worries and eliminate the tendency to repeat operations. This is important not only from the point of view of system efficiency and staff confidence but from users' trust as well. Readers must not be suspicious of the computer's ability to produce correct overdue notices, for example, or they will question the validity of any transaction. Finer (1982) notes that the librarian is under a professional obligation to defend the system against criticism, but sometimes there is a case for a less formal attitude. This is described by Loughridge (1979, p. 117) as 'sympathetic communion with the reader and mutual deprecation of the computer' and can, at appropriate moments, do more for public relations than stoical allegiance.

Another matter, potentially more sensitive, is how much to teach staff of the overall system and its capabilities. One of the advantages of an integrated library system is the single record, accessible from any point. Instruction should be given on how particular operations fit into the whole scheme, and in what ways they interrelate. Files can always be protected by the use of passwords. Whatever the individual library's stance on this it must be remembered that employees who are taught to use the system effectively become more valuable employees in consequence.

Part of the training sessions must be devoted to familiarizing people with unknown equipment. Knowledge should not be assumed, even of the layout of a keyboard. The various hardware and software components should be identified and some explanation made of what they can do. Operators should at least know how to turn their terminals on and off and how to adjust and clean them. This not only reduces the likelihood of any damage to the equipment through misuse but again gives staff a sense of personal control. Health and safety issues likewise must not be bypassed. Scalet (1987) notes that tuition should be given on how to adjust chairs and other workstation furniture. There is little sense in investing in well designed equipment if people are not shown how to use it to their best advantage. Cross and Bawden (1987) favour staff guidance on posture and the best positioning of VDU equipment on the working surface. Library policy on health-related factors such as rest breaks, eyesight testing and pregnancy transfer should be well advertised. Where people are comfortable at work they will be better able to concentrate on the task in hand. Scalet (1987) observes a trend in the modern office of mutual concern between management and employees to reduce absenteeism and accidents through education (a concept initiated in industry). The partnership is just as vital in libraries.

Some understanding of the Data Protection Act must be instilled (Elbra, 1984). Staff should know what details about readers, if any, they are allowed to disclose to others.

Last, but no means least, procedures should be taught on how to cope with systems' failures and breakdowns. In this eventuality, the employee must know how to ensure that service can be maintained and the most appropriate contact to inform of the problem. He must also be aware of how to catch up once things are running normally again. Set formulae will lessen any sense of impotence in the face of machine malfunction. To summarize, then, Damodaran *et al.* (1980) argue that most users will need to learn:

- what the system can do for them;
- how to use the system;
- how to use the equipment;
- how to integrate the computer activities into their existing work roles;
- how to extricate themselves when things go wrong.

Method

How the training programme is run will be dictated by time, available resources, and the functions to be taught. In large organizations, a structured approach is necessary in order to make quite sure that everyone receives instruction. A variety of techniques are available. Using a combination of several is advised since different methods are more suitable for different aspects of information. Matthews (1980) also recommends that a series of training sessions should take place rather than just one; knowledge gained from one session can then be reinforced and consolidated at the next. Training can be carried out through:

- lectures;
- classroom sessions;
- audio-visual presentations;
- demonstrations/visits;
- computer-assisted learning (CAL).

The list is by no means comprehensive but whichever methods are chosen it is important that the trainee has an opportunity to ask questions and clear up any queries he may have. Practical experience should form part of the module wherever possible, and some kind of assessment of the trainee's progress made.

Lectures

Lectures are a useful means of disseminating general information to a wide audience. The drawback is an administrative one – scheduling them at a convenient time when a worthwhile body of staff can attend. If the library already opens late one day a week for a regular training slot this could be used.

Classroom sessions

One of the best ways of giving intensive training is through a classroom type situation (although care should be taken to maintain a relaxed, informal atmosphere). Staff can receive individual attention and the 'hands-on' experience which is so vital to give real meaning to the exercise. There should be a one-to-one ratio of people to terminals. If there is not enough equipment, the value of the sessions is limited. Dyson (1986) says that there should also be one trainer per three trainees. The trainers should be patient and sympathetic, providing instant feedback on employees' progress. Functions should be taught in ascending degree of difficulty with a reminder of why it was felt necessary to automate each process. Relating each action back to the old system may help since learning proceeds best from the familiar. Changing from one automated system to another carries its own peculiar problems: Lefebvre (1985) notes that staff then have to 'unlearn' techniques. Sessions should be broken up, detailed instruction interspersed with group discussion and role play. Recent research by Bouzid and Crawshaw (1987) has confirmed the superiority of distributed wordprocessor training over 'massed' training: this is an axiom relevant to libraries. It should be stressed that proficiency should not be expected immediately.

Audio-visual presentations

Audio-visual presentations are an effective method of teaching but again can only serve as an introduction. However, they give no opportunity for practice. They can be useful for subjects like health and safety for which films, videos and tape slides are available (see Appendix).

Demonstration/visits

Demonstrations or visits to a vendor or another library allow many staff to see the system. Sykes' account (1986), however, highlights the problems. Interviewees complained that they had only been allowed brief access to the terminals, and the large size of the groups deterred nervous people from trying out the system. They did not want to make mistakes in front of others. Leaving staff alone to 'play' with the computer afterwards can however be very beneficial.

Computer-assisted learning

Computer-assisted learning is another possibility (Clayton, 1987). This technique permits people to proceed at their own pace: mistakes are witnessed only by the machine and the individual. The drawbacks are self-evident in the title. Those who are nervous about automation will not feel at home with this medium. Also it may be difficult or costly to obtain appropriate software.

Support mechanisms

Simulation techniques and training sessions can only go part-way towards preparing staff for 'going live'. The real problems and difficulties cannot be envisaged. Gratton (1983) makes this point with some feeling. Practice cannot be given on the scale of transactions that staff will be dealing with, the speed with which they will have to respond, or the pressure from readers. Mechanical failures and breakdowns can only be discussed in the abstract. He says that there is much learning to be done on the day. There are two major sources of support to help meet short-term user needs, human and documentary.

Human support

During the transition period, ease of access to human support is crucial. When problems develop, people tend instinctively to turn to others for guidance and advice. Local experts, supervisors, and computer experts are all much in demand.

A primary source of support is an employee's fellow workers: staff generally ask each other about queries concerning the computer as they do on other routine matters. Questions can often be resolved by one member of the workgroup, an unofficial leader or 'local expert' as Damodaran *et al.* (1980) call him. This person does not necessarily have a higher technical knowledge than the other members of staff but has absorbed more information overall and has taken a greater interest in the implementation of the system. A humane local expert can be indispensable to the organization. One of his greatest assets is accessibility; he tends to be around when things go wrong and he can also answer enquirers on their own level. Prompts can be given on everyday procedures such as how to log on and off and the meaning of certain messages. Although the position of local expert has no formal standing, managers should recognize the valuable role they fill and give concrete acknowledgement wherever possible. His goodwill is worth cultivating.

When local experts cannot provide information or are themselves in doubt, staff look to their immediate supervisors or managers for help. Consequently, managers must know more about the system than their employees, which confirms the need for training senior staff first. For the sake of their own esteem, managers must be able to provide solutions at least part of the time. They must also be approachable in themselves, so that staff will confront them with their problems.

In the event of major systems' breakdown, computer personnel have to be consulted. The amount of computer expertise readily available will vary greatly between libraries and depend on the systems that have been implemented. Where computer personnel are not intrinsic to the library, certain dilemmas may present themselves. A central difficulty is communication. Librarian and technician may appear to speak in different languages. Should the librarian have to seek advice over the telephone he may have problems making the crisis understood. Similarly, if this obstacle is overcome, he may very well find the diagnosis incomprehensible. This particular dilemma may disappear over a period of time as each becomes used to the idiosyncracies of the other, but only if regular contact between the same people is maintained.

The librarian, experienced in his own field and used to resolving problems for himself, may also hesitate to contact systems' experts over every detail. If an identical fault has occurred before, he may feel that he should know the procedure to rectify it. This sort of unease should be avoided and a good relationship established. Library staff must feel able to call in outside help whenever it is needed and, if brought in early, computer personnel may find it easier to deal with the problem. Coaching can be given as the fault is corrected: there is little profit to be gained from a 'magical' resolution with no attendant explanation.

Although there may be a predetermined sequence of people to consult in the event of a computer fault, Gratton (1983, p. 19) argues that staff will still call upon those they know or someone who has proved helpful in the past: this is human nature. He also notes that the urge to call on others or to declare an emergency fades as familiarity with the system grows. Rational decisions replace a kind of crisis management:

> In essence, training, pre- and post-implementation, proved to have more to do with creating self-confidence than skill, although the latter clearly influences self-confidence.

Human support is of great value to the smooth running of a system and the promotion of a healthy atmosphere within the library. But, by its very nature, human support can be erratic and unreliable: employees can leave organizations, or be absent on business and through illness. Fine (1986) has remarked that when staff are experiencing the greatest stress of change it is not unusual for personnel resources to be overextended and withdrawn. There is a need, therefore, for accompanying documentary support.

Documentary support

Damodaran (1980, p. 124) draws up a list of five documents that are required to accompany any item of computer equipment:

- installation instructions;

- maintenance instructions;
- technical specification of the equipment;
- users' guide to operating the equipment;
- users' guide to operating the system.

The first three documents are of limited interest after installation has been completed. However, they should all be retained in a safe place: where systems have been adapted to meet local requirements and are therefore unique, the documentation is extremely valuable. The most essential piece of documentation from the staff point of view is the operator guide – this is frequently supplemented (if not replaced) by the organization's own manual.

Documentary support can prove invaluable in many instances. Even extensive training cannot prepare staff for all the situations they are bound to meet and they are unlikely to retain the finer details in any case. Since libraries are generally open long hours and staffed by people who only have intermittent contact with the system, a written instruction can provide a vital support link for them. Staff turnover, either that within departments or that of new employees joining the organization, is also inevitable. It is rarely cost-effective to hold major training sessions for them and documentation can be used as an aid to learning. Documentation can also serve as a tool to further self-development and as a reminder of procedures. Tangible instructions can be reassuring when human support is unavailable.

In order for the documentation to fulfil these multifarious purposes, however, much effort has to be put into the preparation of the literature. Handing over a manual provided by the vendor may induce panic and an unnecessary sense of insecurity. Sykes (1986) comments that annotations are frequently made, 'bibliographic query' being replaced by the straightforward phrase 'find a book' and 'patron function' amended to the self-explanatory 'who has what out'. In-house production is strongly recommended if time permits.

The effective staff manual should satisfy several basic criteria, simplicity and clarity in presentation being high on the list. Burton and Petrie (1986) advise that one person should actually write the manual, although many can be involved in draft copies and suggestions. This will give uniformity to the text and so facilitate use. Information should be concise and a suitably easy indexing system employed. There is little point in providing documentary support if its very appearance deters use. Where the system is complex, there is usually a need for several different manuals to deal with the various circulation, acquisitions and cataloguing modules. One compendious volume might be offputting, although an authoritative copy should be kept. Clayton (1987) suggests that a shortened, quick reference version is useful, especially for those starting off in a pressurized position at the desk. The manual will need to be updated in any case and attention should be paid to its physical format to make sure this can be done readily: a simple ring-binder file could be used.

Manuals should be available to accompany initial training sessions and staff guided in their application. They will then have an idea of where to look when problems occur. Maximum use should be made of the documentation at these sessions. Staff who come to the manuals 'cold' when they need assistance are far less likely to derive much help from them. The meaning of new terms and phrases can also be covered in systems' documentation.

Detailed operational procedures should be set out in manuals and minute, logical instructions given. Examples should be relevant to the particular library: manuals can thus be used to establish library policy. Lovecy (1984) says that when the computer offers a choice of responses there should be written advice on how to reach decisions. He gives the common example of a reader wishing to exceed his borrower limit. A further advantage of this is to ensure the library actually does have a policy on which line to take. Clayton (1987) remarks that it can promote consistency of practice. This helps to prevent the situation where readers feel victimized if they receive conflicting responses from different assistants. It is particularly important that standards are maintained between separate branches of a single library.

Just as the system changes, the manual should change too. The manual produced before implementation is far from being the definitive version, especially when it is based upon theoretical problems. A revised edition should reflect the concerns of those who have had to use the system. Instruction may, for example, be necessary on how to hold a light pen, or a better way of doing things may become apparent in practice. Burton and Petrie (1986) stress the importance of feedback mechanisms to make sure that these amendments are incorporated. An out-of-date manual is virtually useless and can be very misleading. When system changes are made there is a tendency amongst suppliers to neglect a similar update of the manual. Where a library relies on documentation from a supplier this could be a serious problem.

Damodaran *et al.* (1980) mention the possibility of another type of aid – computer-based documentation. Essentially similar to paper-based documentation, it is available online and has all the advantages of ease of update associated with electronic publishing. Such documentation must however be easy to access and not interrupt normal work – a window within the screen may be a suitable format. Where it is used purely as a learning medium there must be enough terminals to cater for this. Many computer systems have built-in 'help' facilities. These can guide the user when he is unsure what to do next or can give him explanation of abbreviations or codes being used.

It has been shown that training and documentation have a vital link but Baddeley (1988, p. 79) warns that they 'are not substitutes for each other. A good training course does not eliminate the need for well-prepared documentation, and reprinted course notes are not a user guide.'

Over a period of time, staff will become less reliant on manuals, but they do have a continuous purpose and should not be abandoned. They can be used for reference and in the training of new staff.

Further training

The effectiveness of any training programme needs to be assessed and the need for further training identified. One method of assessment is an anonymous questionnaire, completed at the end of the training session. This, however, only judges the relevance of the training session to attenders at that time. Other needs may become apparent as operators become more proficient with the system. The staff of one library expressed regret that they had consumed the total number of days' training offered by the supplier shortly after implementation. In retrospect, they felt that they would have benefited by waiting until they had become more familiar with the software. Additional training sessions are therefore necessary (Lowry, 1985, p. 29):

> Procedures should be established for the continuing training of staff after the implementation of technology commences – after which point training and re-training will never cease.

Refresher courses should be arranged for all staff from time to time. This is important in academic libraries, particularly where staff employed during term-time only have long breaks away from the system. It is particularly vital to remind staff of what to do in the event of system's failure: the fact that a computer is reliable will mean that this eventuality is met infrequently. Comprehensive records should be kept of who has been trained in what and when.

New staff have particular needs, especially junior members joining just before automation takes place. Everything will be strange, not only the people and place but possibly library procedures as well. The newcomer is likely to have missed out on some of the general orientation lectures, and it is unfortunately also true that other employees may not be as cooperative as normal if they are experiencing their own anxieties.

New staff joining after implementation are often trained by their peers. Graham (1987) can see disadvantages in this. Although the library assistant may be extremely good at his job, that does not automatically qualify him for communicating his skill to others. Indeed, proficiency may in itself be a barrier: the trainer may be unable to envisage someone else's problems or remember his own hesitant start. Technicalities can be glossed over and the trainer's own particular method of performing transactions (rather than the standard approach) perpetuated. New employees may also miss out on instruction about adjustable furniture and terminals, which may have become second nature to more established staff.

Training the user

The ultimate beneficiaries of a new automated library system should be the library's users. Training the user is partly realized through contact with the staff, and partly through good publicity. It is to be hoped that well trained employees communicate their positive feelings for the system to the readers. Timely publicity and information also prepares them for automation.

One way of directly involving the public is that reported by Clarke (1985) at Hillingdon. When enrolling new members into the library, both VDU operator and user sit down together at the terminal. The user can then see what details are being put onto the computer and can correct them as necessary. This helps to accustom the user to the machine, is beneficial for public relations, and helps to reduce any feelings of anxiety the user may have towards 'secret data' being kept about him. When dealing with any reader query it is useful for VDUs to be mounted on swivel stands or angled so that assistant and user can view the screen alike without either straining to see, but it is important not to allow other readers in the queue to be able to read details about the reader at the desk.

The OPAC is where the library user will come into closest contact with the computer. As Bichteler (1987) points out, most users lack understanding of the way in which information is organized and accessed: they are happy when they retrieve something, but are unaware of what they fail to find. They may believe that the computer has all the answers or, alternatively, may be discouraged by a failure to retrieve an item they believe to be in the system. They tend to use incorrect Boolean logic and spelling errors are common. Bichteler argues that it is therefore necessary to instruct readers in the principles of searching and the construction of bibliographic records, not just in how to input a search.

Depending on the type of library and on staff availability, a variety of training methods may be used. The most common is to provide a notice and/or leaflet giving brief details about the system and how to use it effectively. This may be supplemented by a manual for more serious searchers. User education sessions may also incorporate something on the use of the computer system and, in some libraries, one-to-one point-of-need training may be appropriate. Bichteler suggests the use of search logs to monitor errors and to adjust the instruction accordingly.

APPENDIX: Some useful addresses

Professional, research and advisory bodies: general

The British Psychological Society
St Andrew House
Princess Road
Leicester
Tel: (0533) 549568

Its members may include workers on the cognitive aspects of human computer interaction who may be contacted via the society.

British Safety Council
62 Chancellors Road
London W6
Tel: (01) 741-1231 or 2371
or (01) 748-7480

Issues publications and runs courses, some of which lead to the Diploma in Safety Management.

Chartered Institute of Building Services
222 Balham High Road
London SW12 9BS
Tel: (01) 675-5211

Publishers of Technical Memoranda TM6 – 'Lighting for Visual Display Units'.

The Division of Ergonomics Practitioners
Scicon Consultancy International Ltd
Sanderson House
49 Berners Street
London W1P 4AQ
Tel: (01) 580-5599

Ergonomics Information Analysis Centre
Dept. of Engineering Production
University of Birmingham
Edgbaston
PO Box 363
Birmingham B15 2TT
Tel: (021) 414-3344

Offers information on human factors.

The Ergonomics Society
c/o Department of Human Sciences
University of Technology
Loughborough
Leicestershire LE11 3TU
Tel: (0509) 263171

The main professional body for ergonomists in the U.K.

HUSAT (Human Sciences and Advanced Technology)
The Elms
Elm Grove
Loughborough
Leicestershire LE11 1RG
Tel: (0509) 212041

Researchers, publishes and undertakes consultancy in all areas of Human Computer Interaction. Have produced a tape/slide presentation on the health and safety aspects of VDUs in conjunction with Loughborough University's Audio Visual Services Department.

The Human Factors Society Inc.
1124 Montana Ave.
Suite B
Santa Monica
California 90406 USA

Humane Technology
PO Box 2
Quorn
Leicestershire LE12 8EG

Runs seminars, courses and sponsors conferences on alleged health hazards of VDU work.

The Institution of Electrical Engineers
Savoy Place
London WC2R 0BL
Tel: (01) 240-1871

May offer information on the human aspects of I.T.

London Hazards Centre
3rd Floor
Headland House
308 Grays Inn Road
London WC1X 8DS
Tel: (01) 837-5605

Publishes 'VDU Hazards Handbook' and 'Repetition Strain Injuries' and gives advice.

NKR Environment Ltd
73 Welbeck Street
London W1M 7HA
Tel: (01) 486-7051

The National Computing Centre Ltd
Oxford Road
Manchester M1 7ED
Tel: (061) 228-6333

Advises and publishes on the ergonomic considerations of VDU installations.

National Electronics Council
Savoy Hill
London WC2
Tel: (01) 836-4264

ROSPA Film Library
Cannon House
The Priory Queensway
Birmingham B4 6BS
Tel: (021) 200-24101

Have 'The VDU Film' for hire which takes a positive look at the correct positioning of equipment for operator comfort, deals with maintenance and discusses some of the alleged health risks.

Space Planning Services Ltd
Western House
Uxbridge Road
Hillingdon
Middlesex UB10 0LY
Tel: (01) 573-2271

Independent consultancy expressly concerned with providing working environments that make a positive contribution to business operations.

System Concepts Ltd
Museum House
Museum Street
London WC1A 1JT
Tel: (01) 636-5912

Independent consultants specialising in systems.

Medical associations and self-help groups

Association of Optical Practitioners (Optometrists)
233 Blackfriars Road
London SE1 8NW
Tel: (01) 261-9661

Produces booklets 'VDUs on Site', 'Vision and VDUs' and 'VDUs and You: Information for Operators'.

Eye Care Information Bureau
4 Ching Court
Shelton Street
London WC2H 9DG
Tel: (01) 836-1765

Welcomes telephone and written enquiries on all aspects of eye care.

Dr V. M. Reading
Department of Visual Science
Institute of Ophthalmology
University of London
Judd Street
London WC1H 9QS
Tel: (01) 387-9621

Medical Research Council
20 Park Crescent
London W1N 4AL
Tel: (01) 636-5422

Government body which can give information on research in medical matters.

Optical Information Council
Walter House
418/422 Strand
London WC2R 0PB
Tel: (01) 353-3556

RSI Support Group
Hillingdon Association of Voluntary Services
Volunteer Bureau
Christ Church
Redford Way
Uxbridge
Middlesex UB8 1SZ
Tel: (0895) 39830 or 59806

Telephone help line and information for RSI sufferers.

Employment advice

BIFU (Banking Insurance and Finance Union)
Sheffield House
17 Hillside
Wimbledon
London SW19 4NJ
Tel: (01) 946-9151

Publish guidelines on work with VDUs.

Department of Trade and Industry
Information Technology Division
1–19 Victoria Street
London SW1
Tel: (01) 215-7877

May give advice on human aspects of IT.

Employer's Medical Advisory Service
See **Health and Safety Executive**

Health and Safety Executive
Broad Lane
Sheffield S3 7HQ
Tel: (0742) 78141

Offers general advice, publishes books and leaflets and supports HSELINE, a publicly available database. Incorporates Employer's Medical Advisory Service.

The Industrial Society
Peter Runge House
3 Carlton House Terrace
London SW1Y 5DG
Tel: (01) 839-4300

Carries out surveys into working conditions associated with work done on VDUs for organizations.

Industrial Alcoholism Unit
82 West Regent Street
Glasgow G2 2QF
Tel: (041) 332-7936

If things get really rough!

Information Advice and Resources for Office Workers
See **VDU Workers' Rights Campaign.**

Industrial Relations Services
67 Maygrove Road
London NW6 2EJ
Tel: (01) 328-4751

Institution of Occupational Safety and Health
222 Uppingham Road
Leicester LE5 0QG
Tel: (0533) 768424

Labour Research Dept.
78 Blackfriars Road
London SE1 8HF
Tel: (01) 928-3649

Publishes *VDUs: Health and Jobs.*

London Employee Alcoholism Project
c/o Rev Francis Jakeman
Industrial Chaplain
128 Pinner View
Harrow
Middlesex
Tel: (01) 427-8678

TUC Centenary Institute of Occupational Health
London School of Hygiene and Tropical Medicine
Keppel Street
London WC1
Tel: (01) 580-2386

Runs MSc courses in occupational health and occupational medicine and short courses. Provides advice to industry and the TUC, and carries out research.

VDU Workers' Rights Campaign
c/o City Centre Project
2nd Floor
Sophia House
32 Featherstone Street
London WC1Y 8QX
Tel: (01) 608-1338

Issues general advice and leaflets on work with VDUs.

Work Research Unit – Advisory, Conciliation and Arbitration Service
St Vincent House
30 Orange Street
London WC2H 7HH
Tel: (01) 839-9281/9289

May give advice on work-related issues and publish bibliographies and papers.

Trade unions offering information

APEX: Association of Professional, Executive and Computer Staff
22 Worple Road
London SW19 4DF
Tel: (01) 947-3131

Publishes *New Technology: A Health and Safety Report* and *Job Design and New Technology*.

AUEW, Amalgamated Union of Engineering Workers
Onslow Hall
Little Green
Richmond
Surrey TW9 1QN
Tel: (01) 948-2271

CPSA, Civil and Public Services Association
215 Balham High Road
London SW17 7QB
Tel: (01) 672-1299

NALGO, National and Local Government Officers Association
1 Mabledon Place,
London WC1H 9AJ
Tel: (01) 388-2366

Equipment suppliers

Acco Europe Ltd
Halesowen Industrial Park
Hereward Rise
Halesowen
West Midlands
B62 8AN
Tel: (021) 550-8883

Provides a range of computer accessories such as tilt/swivel stands, copy holders, filter screens and acoustic hoods. Also suppliers of system furniture.

BT Buylines (Sales Office)
10–18 Manor Gardens
London N7 6JY
Tel: (01) 281-4422

Supplies furniture, including adjustable desks.

Balmforth Engineering Ltd
Finway
Library Systems Division
Dallow Road
Luton LU1 1TE
Tel: (0582) 31171

Suppliers of library furniture.

Barratt Officentres
136 Lawley Street
Birmingham B4 7XY
Tel: (021) 359-7701

Suppliers of office furniture and lighting.

Blyde-Barton Furniture Ltd
Reepham Road
Norwich NR6 6NA
Tel: (0603) 789000

Furniture suppliers.

Business Seating (Midlands) Ltd
Unit 4
Stechford Trading Estate
Lyndon Road
Stechford
Birmingham B33 8BR
Tel: (021) 783-9810

Provides a range of chairs and other office furniture.

EDL Industries Ltd (Lighting Division)
Redbrook Lane
Brereton
Rugeley
Staffordshire WS15 1QY
Tel: (08894) 2112

Designers and manufacturers of lighting fittings and cable supports.

Gaylord Bros Inc.
PO Box 4901
Syracuse
NY 13221
USA

Suppliers of the Computer Caroussel Work Centre to hold six OPAC terminals.

Giroflex Ltd
6 Giltspur Street
London EC1A 9DE
Tel: (01) 236-0653

Supplies office seating.

Libraco Ltd
Warspite Road
London SE18 5NX
Tel: (01) 855-6102

Suppliers of standard library reading tables and tailor-made furniture.

Library Design and Engineering Ltd
44 Gloucester Avenue
London NW1 1AA
Tel: (01) 722-0111/3

Suppliers of library shelving, study carrels, desks, chairs and other equipment.

Librex Educational Co
Colwick Road
Nottingham NG2 4BG
Tel: (0602) 504664/580032

Supplies library equipment and furniture.

Mac User Group
55 Linkside Avenue
Oxford OX2 8JE
Tel: (0865) 58027

Distributors of REDIT, a general-purpose resource editor for the software developer which includes the facility to redefine the keyboard to the Dvorak layout.

Misco Computer Supplies Ltd
4 The Western Centre
Western Road
Bracknell
Berkshire RG12 1BR
Tel: (01) 998-9068

Supplies a range of computer equipment including copy holders, anti-glare screens, antistatic mats etc.

PCD Maltron Ltd
15 Orchard Lane
East Molesey
Surrey KT8 0BN
Tel: (01) 398-3265

Suppliers of the Maltron keyboard.

Palantir Ltd
23–24 High Street
Woking
Surrey GU21 1BW
Tel: (0486) 229144

Markets MACTYPE, a typing tutor program for the Apple Macintosh for either the Qwerty or the Dvorak layout.

Paragon Furniture, Inc.
2819 Blystone Lane
Dallas
Texas 75220
USA

Manfacturers of multi-terminal work centres, including the Computer Caroussel Work Centre. Contact for details of suppliers.

Pedley Woodwork Ltd
Shirehill Works
Saffron Walden
Essex CB11 3AL
Tel: (0799) 22461

Suppliers of desks, shelving and other wooden furniture.

President Office Furniture
Lincoln House
Colney Street
St Albans
Hertfordshire AL2 2DX
Tel: (09276) 7211

Furniture suppliers

Project Office Furniture plc
Showroom and Factories
Haverhill
Suffolk CB9 8QJ
Tel: (0440) 705411

Suppliers of office furniture and lighting.

B. Serota Ltd
Acme Works
Rendlesham Road
Clapton
London E5 8PS
Tel: (01) 985-1736/1787

Library furniture suppliers.

Special Systems Industry BV
Lange Voorhout 45a
2514 EC Den Haag
The Netherlands

Manufacturers of the Velotype keyboard

Steelcase Strafor (UK) Ltd
100 Avenue Road
London NW3 3HF
Tel: (01) 586-5933

Suppliers of office furniture.

Terrapin International Ltd
Bond Trading Estate
Bond Avenue
Bletchley
Milton Keynes MK1 1JJ
Tel: (0908) 270900

Specialize in library issue desks and other custom-made furniture.

Vickers plc
Dartford
Kent DA1 1NY
Tel: (0322) 23477

Supply office furniture, including chairs.

Wilkinsons Furniture Ltd
Monkhill
Pontefract
West Yorkshire WF8 2NS
Tel: (0977) 791191

Suppliers' associations

Business Equipment Trade Association
8 Southampton Place
London WC1A 2EF
Tel: (01) 405-6233

Publish *Introducing the Visual Display Unit to the Office*.

European Computer Manufacturers' Association
114 Rue du Rhone
1204 Geneva
Switzerland

Produces leaflet entitled 'Ergonomic Recommendations for VDU Work Places'.

Publishers

Ergolab Ltd
9 Ridgway Road
Farnham
Surrey GU9 8NN
Tel: (0252) 710502

Publications include: *Ergonomic Principles in Office Automation* by T. A. R. Berns and *Information Ergonomics* by T. B. K. Ivegård.

Millbank Films Ltd
Ivybridge House
1 Adam Street
London WC2N 6AW
Tel: (01) 839-7176

Have for sale or hire a film entitled *Somebody told me . . .* about modern technology in the office.

TV Choice Video Productions
80 St Martins Lane
London WC2N 4AA
Tel: (01) 379-0873

Offer a video, *Using Office Automation*.

Taylor and Francis Ltd
4 John Street
London WC1N 2ET
Tel: (01) 405-2237

Publishers of a wide range of books and journals on ergonomics.

John Wiley and Sons Ltd
Baffins Lane
Chichester
Sussex
Tel: (0243) 779777

Publishers of *Ergonomics of VDU Installation* by J. M. Bird, *Health Hazards of VDTs?* by B. G. Pearce, and others.

REFERENCES AND BIBLIOGRAPHY

Abernethy, C. N. and Hodes, D. G. (1987), 'Ergonomically determined pointing device (mouse), design,' *Behaviour and Information Technology*, **6**, **(3)**, 311–14.

Alderfer, C. P. (1969), 'An empirical test of a new theory of human needs'. *Organizational Behaviour and Human Performance*, **4**, 142-75.

Alderfer, C. P. (1972), *Existence, Relatedness and Growth: human needs in organizational settings*, New York: Free Press.

Anderson, J. R. (1980), *Cognitive Psychology and its Implications*, San Francisco, California: Freeman.

Antin, J. F. (1988), 'An empirical comparison of menu selection, command entry and combined modes of computer control', *Behaviour and Information Technology*, **7**, **(2)**, 173–82.

APEX: London and Home Counties Area Technology Subcommittee (1985), *New Technology: a health and safety report*, London: Apex.

Argyle, M. (1972), *The Social Psychology of Work*, Harmondsworth: Penguin.

Atkinson, H. C. (1984), 'Strategies for change: Part II, *Library Journal*, **109**, **(5)**, 556–7.

Baddeley, A. D. and Hitch, G. (1974), *Working Memory* ('Recent advances in learning and motivation', vol. 8), New York: Academic Press.

Baddeley, A. D. and Lieberman, K. (1980), *Spatial Working Memory* ('Attention and performance', vol. 8), Hillsdale, NJ: Erlbaum.

Baddeley, J. (1988), 'Winning ways with users', *Systems International*, April, 77–9.

Bagley, D. E. and Oyston, E. (1982), *Automation in a Polytechnic Library: fifteen years' development at Hatfield*, London: Library Association.

Bammer, G. (1987), 'Repetition strain injuries affect VDU operators too'. *Health and Safety at Work*, February, 39–41.

Barclay, N. L. (1986), 'Breaking the world speed record', *The Guardian*, 27 March, 25.

Bayne, V. J. (1986), 'Paper outlining and experience of the CPSA involvement in job design' in *Job design and VDU work: a one-day conference, 1 October 1986*. Humane Technology, Quorn, 59–73.

Becket, M. and Smith, H. B. (1986), 'Designing a reference station for the information age', *Library Journal*, **111,** 15 April, 42–6.

Beckman, M. (1987), 'Library equipment in a changing environment' in *53rd IFLA Council and General Conference, Brighton, 16–21 August 1987*, 16–2 to 16–3.

Bendix, T. and Jessen, F. (1986), 'Wrist support during typing – a controlled electromyographic study, *Applied Ergonomics*, **17**, **(3)**, 162–8.

Bennett, J. L. (1986), 'Observations on meeting usability goals for software products', *Behaviour and Information Technology*, **5**, **(2)**, 183–93.

Bergqvist, U. (1986), 'The alleged reproductive hazard of VDU work – an overview' in *Allegations of Reproductive Hazards from VDUs: the second international scientific meeting, London, 25 and 26 November 1986*, Humane Technology, Quorn, 3–14.

Bichteler, J. (1986), 'Human aspects of high tech in special libraries', *Special Libraries*, **77**, **(3)**, 121–8.

Bichteler, J. (1987), 'Technostress in libraries: causes, effects and solutions', *Electronic Library*, **5**, **(5)**, 282–7.

Blair, L. (1986), 'Is it just a VDU scare?', *Business Systems and Equipment*, February, 45, 47.

Blauner, R. (1964), *Alienation and Freedom: the factory worker and his industry*, Chicago: University of Chicago Press.

Borgman, C. L. (1988), 'Human factors in the use of information systems: research methods and results' in N. Tudor-Šilović and I. Mihel (eds), 'Information research: Research methods in library and information science, *Proceedings of an International Seminar on Information Research, Dubrovnik, Yugoslavia, 19–24 May, 1986*, Taylor Graham, 139–65.

Boss, R. W. (1984), *The Library Manager's Guide to Automation*, (2nd edn.) White Plains, NY: Knowledge Industry Publications.

Bouzid, N. and Crawshaw, C. M. (1987), 'Massed versus distributed wordprocessor training', *Applied Ergonomics*, **18**, **(3)**, 220–2.

Bower, G. H. and Reitman, J. S. (1972), 'Mnemonic elaboration in multilist learning, *Journal of Verbal Learning and Verbal Behaviour*, **253**, 478–85.

Boyle, H. M. (1986), 'VDUs and Health', A letter from Mr H. M. Boyle, British Embassy, Paris to Mr C. Barrow, Alvey Directorate, 9 April.

BP Group Occupational Health Centre (1987), *Visual Display Units (VDUs): management guideline on health aspects of work with visual display units*, Guildford: BP Group.

Braidwood, P. (1988), 'VDUs: are they safe in pregnancy? *Observer*, 7 August, 35.

Brandt, L. (1986), 'The effect of VDT use on pregnancy outcome – an epidemiological study of the potential hazard on pregnancy, working with VDTs in *Allegations of Reproductive Hazards from VDUs: the second international scientific meeting, London, 25 and 26 November, 1986*. Humane Technology, Quorn, 141–4.

Brix, K. A. and Butler, W. J. (1986), 'A review of proposed, current and completed epidemiological studies of VDU workers and pregnancy outcomes in *Allegations of Reproductive Hazards from VDUs: the second international scientific meeting, London, 25 and 26 November, 1986*, Humane Technology, Quorn, 31–50.

Brown, I. D. (1965), 'A comparison of two subsidiary tasks used to measure fatigue in car drivers, *Ergonomics*, **8**, 467–73.

Brown, I. D. and Poulton, E. C. (1961), 'Measuring the 'spare mental capacity' of car drivers by a subsidiary task', *Ergonomics*, **5**, 35–40.

Brownrigg, E. B. (1982), 'Library automation: building and equipment considerations in implementing computer technology in G. B. McCabe *et al.* (eds), *Advances in Library Administration and Organization*, **1**, London: JAI Press, 43–53.

Buesen, J. (1984), 'Product development of an ergonomic keyboard, *Behaviour and Information Technology*, **3**, **(4)**, 387–90.

Burton, P. (1987), 'Microcomputers as innovation: policies for implementation in libraries and information services, *Electronic library*, **5**, **(4)**, 210–20.

Burton, P. F. and Petrie, J. H. (1986), *The Librarian's Guide to Microcomputers for Information Management*, Wokingham, Berks: Van Nostrand Reinhold (UK).

Busch, J. (1984), 'Got the VDTs?', *Small Computers in Libraries*, **4**, **(5)**, 7.

Butler, J. and Brix, K. A. (1986), 'Video display terminal work and pregnancy outcome in Michigan clerical workers in *Allegations of Reproductive Hazards from VDUs: the second international scientific meeting, London, 25 and 26 November*, Humane Technology, Quorn, 67–91.

Buxton, A. and Trenner, L. (1987), 'An experiment to assess the friendliness of error messages from interactive information retrieval systems', *Journal of Information Science*, **13, (4)**, 197–209.

Cakir, A., Hart, D. J. and Stewart, T. F. M. (1980), *Visual Display Terminals: a manual covering ergonomics, workspace design, health and safety, task organisation*, Chichester: Wiley.

Carroll, J. M. and Rosson, M. B. (1987), 'Paradox of the active user' in J. M. Carroll (ed.), *Interfacing Thought: Cognitive aspects of human–computer interaction*, Cambridge, Mass: MIT Press, 80–111.

Chapanis, A (1960), 'Human engineering' in C. D. Flagle, W. H. Higgins and R. N. Roy (eds), *Operations Research and Systems Engineering*, Baltimore: The Johns Hopkins University Press, 534–82.

(The) Civil and Public Services Association Research Department (1983), *New Technology Bulletin*, September.

Clarke, J. E. (1985), 'The human/IT interface: a librarian's viewpoint', *Library and Information Research News*, **8, (30)**, 5–9.

Clayton, M. (1987), *Managing Library Automation*, Aldershot: Gower.

Cluff, E. (1985), 'Examining the health hazards of the VDU', *Computing: The Magazine*, 27 June, 30.

Cole, I., Lansdale, M. and Christie, B. (1985), 'Dialogue design guidelines' in B. Christie (ed.), *Human Factors of Information Technology in the Office*, Chichester: Wiley, 212–41.

Coleman, V. (1978), *Stress Control*, London: Pan.

Collins Dictionary of the English Language (1979), London: Collins.

(The) Concise Oxford Dictionary (1975), (5th edn.), Oxford: Clarendon Press.

Conrad, R. (1964), 'Acoustic comparisons in immediate memory', *British Journal of Psychology*, **55**, 75–84.

Conyers, A. (1987), 'Staff training and effectiveness' in D. Revill (ed.), *Personnel Management in Polytechnic Libraries*, Aldershot: Gower, 53–75.

Corbin, J. (1985), *Managing the Library Automation Project*, Phoenix, Arizona: Oryx Press.

Corlett, N., Wilson, J. and Manenica, I. (eds), (1986), *The Ergonomics of Working Postures: models, methods and cases. The proceedings of the first international occupational ergonomics symposium, Zadar, Yugoslavia, April 1985*, London: Taylor and Francis.

Cross, R. and Bawden, D. (1987), 'Information technology: human and organizational factors', *Journal of Information Science*, **13**, 277–84.

Cushman, W. H. (1986), 'Reading from microfiche, a VDT, and the printed page: subjective fatigue and performance, *Human Factors*, **28, (1)**, 63–73.

Czaja, S. J., Cary, J. M., Drury, C. G. and Cohen, B. G. (1987), 'An ergonomic evaluation of traditional and automated office environments', *Office Technology and People*, **3, (3)**, 231–46.

Dakshinamurti, G. (1985), 'Automation's effect on library personnel', *Canadian Library Journal*, **42, (6)**, 343–51.

Damodaran, L., Simpson, A. and Wilson, P. (1980), *Designing Systems for People*, Manchester: NCC Publications.

Damodaran, L. (1984), 'Introduction' in B. G. Pearce (ed.), *Health Hazards of VDUs?*, Chichester, Wiley, 3–4.

Damodaran, L. and Ip, K. (1986). 'The job design perspective in information technology' in *Job Design and VDU Work: a one day conference, 1 October 1986*, Humane Technology, Quorn, 19–30.

Danks, D. (1984), 'Persuading users of usability's value', *Computing*, 31 May, 14–15.

De Jong, J. R. and Roggeveen, C. (1969), 'The sitting employee and the standing customer', *Ergonomics*, **12, (2)**, 348–55.

Dean, S. C. (1987), *Durham County Library: annual report, 1986–87*, Durham: County Library.

Deese, J. (1964), *Principles of Psychology*, Boston, Mass: Allyn and Bacon.

'Defence system designers urged to improve displays' (1985), *Computing: The Newspaper*, 18 July, 6.

Dickson, J. (1984), 'An analysis of user errors in searching an online catalog', *Cataloguing and Classification Quarterly*, **4, (3)**, 1–17.

Doyle, C. (1988), 'Stress that stares us in the face', *Daily Telegraph*, 26 July, 11.

Dressel, D. L. and Francis, J. (1987), 'Office productivity: contributions of the workstation', *Behaviour and Information Technology*, **6, (3)**, 279–84.

Dunsmore, H. E. 'Using formal grammars to predict the most useful characteristics of interactive systems', *Office automation conference digest*, San Francisco, April 5–7, 1982, 53–56.

Dwyer, G. (1984), 'Evaluating online public access catalogues: a checklist', *State Librarian*, **32, (3)**, 42–3.

Dyson, J. (1986), 'Training for specific skills – automation' in R. Prytherch (ed.), *Handbook of Library Training Practice*, Aldershot: Gower, 277–321.

Elbra, R. A. (1984). *Guide to the Data Protection Act*, Manchester: NCC Publications.

Electronic Library (1985), **3, (1)**, 15.

Else, L. (1985), 'VDUs: a risk to health?', *Computing: The Newspaper*, 21 March, 25.

Ericsson Information Systems (1983), *Ergonomic Principles in Office Automation: state of the art reports and guidelines on human factors in the office environment*, Bromma, Sweden: Ericsson Information Systems AB.

Fayen, E. G. (1983), *The Online Catalog: improving public access to library materials*, White Plains, NY: Knowledge Industry Publications.

Ferris, A. (1987), 'Better by design', *Systems International*, December, 27, 28, 30, 32.

Fine, S. (1982), 'Human factors and human consequences: opening commentary' in A. Kent and T. J. Galvin (eds.), *Information Technology: critical choices for library decision makers*, New York: Marcel Dekker, 209–24.

Fine, S. F. (1986), 'Technological innovation, diffusion and resistance: An historical perspective', *Journal of Library Administration*, **7, (1)**, 83–108.

Finer, R. (1982), 'The human side of change' in *CLAIM Report no. 14*, Loughborough University, Loughborough: Centre for Library and Information Management, 31–49.

Fleischer, A. G., Rademacher, U. and Windberg, H.-J. (1987), 'Individual characteristics of sitting behaviour', *Ergonomics*, **30, (4)**, 703–9.

Fletcher, J. (nd), *Libraries and Automated Systems: a comparison of systems and costs*, Bangor: Institute of Economic Research, University College of North Wales.

Fosdal, J., (1985), 'Upper limb disorders among keyboard users in Britain: a union view', in *Alleged Health Hazards of Work at VDUs: current industrial relations issues. A conference held by Humane Technology in London on 26 and 27 November, 1985*, Humane Technology, Quorn, 111–29.

Fox, J. G. (1983), 'Industrial music' in D. J. Oborne and M. M. Gruneberg (eds), *The Physical Environment at Work*, Chichester: Wiley, 211–26.

Freedman, M. J. (1981), 'Circulation systems past and present', *Journal of Library Automation*, **14, (4)**, 278–85.

Gage, N. and Berliner, D. (1979), *Educational Psychology*, Chicago: Rand-McNally.
Galitz, W. O. (1984), *Humanizing Office Automation: the impact of ergonomics on productivity*, Wellesley: QED Information Sciences, Inc.
Germann, M. P., Nowak, E. L. and Wilke, J. S. (1987), 'Form design simplifies online search procedures, *Information Technology and Libraries*, **6, (4)**, 313–16.
Gibson, J. E. (1965), Adaptive and learning control systems' in R. E. Machol (ed.), *System Engineering Handbook*, New York: McGraw-Hill, 30–1 to 30–20.
Gilad, I. and Pollatschek, M. A. (1986), 'Layout simulation for keyboards', *Behaviour and Information Technology*, **5, (3)**, 273–87.
Gorman, M. (1985), *The Impact of Technology on the Organization of Libraries* (CLSI Annual Lectures on Library Automation no. 1), London: CLSI.
Gould, J. D., Alfaro, L., Finn, R., Haupt, B., Minuto, A. and Salaun, J., (1986), 'Why is reading slower from CRT displays than from paper?' in *A Cradle for Human Factors. Proceedings of the Human Factors Society 30th Annual Meeting, Dayton, Ohio, 29 September–3 October 1986, vol. 2*, Santa Monica, Cal., The Human Factors Society, 834–6.
Gould, J. D. and Grischkowsky, N. (1984), 'Doing the same work with paper and cathode ray tube displays (CRT)' in E. Grandjean (ed.), *Ergonomics and Health in Modern Offices*, London: Taylor and Francis, 329–38.
Graham, K. R. (1987), 'The library assistant' in D. Revill (ed.) *Personnel Management in Polytechnic Libraries*, Aldershot: Gower, 142–54.
Grandjean, E. (1980), *Fitting the Task to the Man: an ergonomic approach*, London: Taylor and Francis.
Grandjean, E. (1987), *Ergonomics in Computerized Offices*, London: Taylor and Francis.
Gratton, P. D. (1983), *Automation in Derbyshire County Libraries*, London: Library Association.
(The) Great On-going Ergonomic Chair Survey (1986), *Business Systems and Equipment*, February, 54–5.
Grey, S. M. and Wilson, J. R. (1984), *The Ergonomics of Library Issue Desks*, Report to the British Library Research and Development Department on Project no. SI/G/563, London: British Library.
Hamdy, W. R. (1987). *The Impact of Visual Display Units on Vision*, Master's dissertation, Loughborough: Loughborough University.
Hammond, C. (1986), 'The hazards of VDUs', *Practical Computing*, **9, (6)**, 73–5.
Hampton, F. (1985), 'A sight for sore eyes', *Mind Your Own Business*, June, 24.
Handy, C. B. (1981), *Understanding Organizations*, (2nd edn.), Harmondsworth: Penguin.
Health and Safety Executive (1983), *Visual Display Units*, London: HMSO.
'Health at Work: 1. Pregnancy and VDUs' (1984), *Labour Research*, December, 300.
Hedge, A. (1987), Office health hazards: an annotated bibliography', *Ergonomics*, **30, (5)**, 733–72.
Herring, J. E. and Mackenzie, J. A. (1986), *Planning for Library Automation: Aberdeen City Libraries*, London: Library Association.
Herzberg, F., Mausner, B. and Snyderman, B. (1959), *The Motivation to Work*, New York: Wiley.
Hobday, S. (1985), 'Letting the shape of the keyboard take the strain', *Computing*, 31 October, 33.

Holmes, T. H. and Rahe, H. (1967), 'The social readjustment rating scale', *Journal of Psychosomatic Research*, **11**, 213–18.

Holroyd, K. (1986), 'Job design and new technology – a trade union initiative' in *Job Design and VDU Work: a one day conference, 1 October 1986*, Humane Technology, Quorn, 31–51.

Hughes, G. (1985), 'A large employer's approach to the use of VDUs' in *Alleged Health Hazards of Work at VDUs: current industrial relations issues. A conference held by Humane Technology in London on 26 and 27 November, 1985*, Humane Technology, Quorn, 141–60.

Hunter, I. M. L. (1964), *Memory*, London: Whitefriars Press Ltd.

Hunter, P. (1983), 'Vertical lights do the trick', *Computer Weekly*, 24 November, 20.

Hünting, W. (1984), 'Postural loads at VDT work-stations' in B. G. Pearce (ed.), *Health Hazards of VDTs?* Chichester, Wiley, 119–36.

Ilg, R. (1987), 'Ergonomic keyboard design', *Behaviour and Information Technology*, **6, (3)**, 303–9.

'Irresponsible reporting derides VDU operators' health problems' (1986), *Health and Safety at Work*, **8, (1)**, 46–7.

James, R. (1987), 'Training and the management of automation, *Training and Education*, **4, (3)**, 53–7.

Jeffreys, S. (1984), VDUs: more than meets the eye', *Computer News*, 26 April, 22.

Jones, N. and Jordan, P. (1987), *Staff Management in Library and Information Work*, (2nd edn.), Aldershot: Gower.

Kahn, H. and Cooper, C. L. (1986), 'Computing stress', *Current Psychological Research and Reviews*, Summer, 148–62.

Karim, A. N. (1986). *Human Factors and Circulation Control Systems*, Master's dissertation, Loughborough: Loughborough University.

Klatzky, R. L. (1980), *Human Memory: structures and processes*, San Francisco: Freeman.

Knave, B. G. (1983), 'The visual display unit' in *Ericsson Information Systems, Ergonomic Principles in Office Automation: state of the art reports and guidelines on human factors in the office environment*, Bromma, Sweden: Ericsson Information Systems, 11–41.

Knill-Jones, R. P. (1986), 'Epidemiological approaches to reproductive problems in the workplace' in *Allegations of Reproductive Hazards from VDUs: the second international scientific meeting, London, 25 and 26 November 1986*, Humane Technology, Quorn, 15–30.

Kolers, P., Duchnicky, R. L. and Ferguson, D. C. (1981), 'Eye movement measurement of readability of CRT displays', *Human Factors*, **23**, 517–27.

Kyllonen, P. C. and Alluisi, E. A. (1987), 'Learning and forgetting facts and skills' in G. Salvendy (ed.), *Handbook of Human Factors*, New York: Wiley, 45–71.

Land, T. (1985), 'VDU screens cleared of threat to health', *Financial Times*, 30 May, 15.

Lansdale, M. (1985), 'Beyond dialogue design guidelines: the role of mental models' in B. Christie (ed.), *Human Factors of Information Technology in the Office*, Chichester: Wiley, 242–70.

Lansdown, J. (1984), 'Not only computing – also art', *Computer Bulletin*, September, 10.

Lefebvre, C. J. (1985), *Training for Automation: a study of the introduction of junior staff to automated circulation systems in academic libraries*, Master's dissertation, Loughborough: Loughborough University.

Library and Information Technology Centre (1987), *State of the Art of the Application of New Information Technologies in Libraries and their Impact on Library Functions in the United Kingdom*, London: Library and Information Technology Centre.

London Hazards Centre (1984), *The Hazards of Visual Display Units (VDUs): hazard information pack*, London: London Hazards Centre.

Loughridge, F. B. (1979), 'Some arguments against computers in libraries', *Assistant Librarian*, **72**, 117.

Lovecy, I. (1984), *Automating Library Procedures: a survivor's handbook*, London: Library Association.

Lowry, C. B. (1985), 'Technology for libraries: six rules for management', *Library Hi Tech*, **3, (3)**, 27–9.

Lucas, H. C. (1974), *Towards Creative Systems Design*, Columbia: Columbia University Press.

McDonald, A. D. (1986), 'Work and pregnancy in Montreal: findings on work with Visual Display Units' in *Allegations of Reproductive Hazards from VDUs: the second international scientific meeting, London, 25 and 26 November*, Humane Technology, Quorn, 53–66.

MacKay, C. (1984), 'Visual display unit operation: possible reproductive effects' in *Allegations of Reproductive Hazards from VDUs: proceedings of an international meeting, London, 29–30 November*, Humane Technology, Quorn, 137–59.

MacKay, C. (1985), 'Occupational upper limb disorders – causes and prevention' in *Alleged Health Hazards of Work at VDUs: current industrial relations issues. A conference held by Humane Technology in London on 26 and 27 November 1985*. Humane Technology, Quorn, 77–88.

MacKay, C. and Cox, T. (1984), 'Occupational stress associated with visual display unit operation' in B. G. Pearce (ed.), *Health Hazards of VDTs?*, Chichester: Wiley, 137–43.

MacMorrow, N. (1987), 'Do VDUs make you sick?' *Aslib Proceedings*, **39, (3)**, 65–74.

Mace, C. A. (1968), *The Psychology of Study* (3rd edn.), Harmondsworth: Penguin.

Malinconico, S. M. (1983), 'People and machines: changing relationships?', *Library Journal*, **108, (21)**, 2222–4.

Malone, T. W. and Lepper, M. (1983), *Motivation, Cognition, and Computerized Instruction*, Paper presented at the Office of Naval Research Conference on Aptitude, Learning and Instruction: Cognitive and Affective Process Analysis, Stamford, California.

Mandal, A. C. (1986), 'Investigation of the lumbar flexion of office workers' in N. Corlett, J. Wilson and I. Manenica (eds), *The Ergonomics of Working Postures: models, methods and cases, The Proceedings of the First International Occupational Ergonomics Symposium, Zadar, Yugoslavia, 15–17 April 1985*, London: Taylor and Francis, 345–54.

Mann, P. (1984), 'Under pressure: how can libraries survive and expand?', *Library Association Record*, **86, (4)** 167.

Marchant, M. P. (1982), 'Participative management, job satisfaction, and services', *Library Journal*, **107, (8)**, 782–4.

Marek, T., Noworal, C., Gedliczka, A. and Matuszek, L. (1984), 'A study of a modified VDT-stand arrangement', *Behaviour and Information Technology*, **3, (4)**, 405–9.

Maslow, A. H. (1943), 'A theory of human motivation', *Psychological Review*, **50**, 370–96.

Mason, R. (1984), 'Mason on micros. Ergonomics: the human and the machine', *Library Journal*, 15 February, 331–2.

Matthews, J. R. (1980), *Choosing an Automated Library System: a planning guide*, Chicago: American Library Association.

Matthews, J. R., Lawrence, G. S. and Ferguson, D. K. (1983), *Using Online Catalogs: a nationwide survey*, New York: Neal-Schuman.

Meichenbaum, D. (1977), *Cognitive Behaviour Modification: an integrative approach*, New York, Plenum.

Mild, K. H. (1986), 'Biological effects of low frequency electromagnetic fields' in *Allegations of Reproductive Hazards from VDUs: the second international scientific meeting, London, 25 and 26 November 1986*, Humane Technology, Quorn, 135–40.

Miller, G. A. (1956), 'The magical number seven plus or minus two: some limits on our capacity for processing information', *Psychological Review*, **63**, 81–97.

Misco (1988), *Catalogue of Products*, April–May.

Murrell, K. F. H. (1971), *Ergonomics: man in his working environment*, London: Chapman and Hall.

Nadler, D. A. (1980), *Concepts for the Management of Organizational Change*, London: O. R. & C. Consultants.

National Electronics Council (1983), *Human Factors and Information Technology*, London: National Electronics Council.

Neisser, U. (1967), *Cognitive Psychology*, New York: Appleton-Century-Crofts.

Nelson, B. (1984), 'VDT and workers', *Library Hi Tech News*, **1, (2)**, 1, 9, 10.

Newmark, M. E. and Penry, J. K. (1979), *Photosensitive Epilepsy: a review*, New York: Raven Press.

Nielsen, J. (1987), 'Classification of dialog techniques: A CHI +GI '87 workshop, Toronto, 6 April, *SIGCHI Bulletin*, **19, (2)**, 30–35.

Norman, D. A. and Fisher, D. (1982), 'Why alphabetic keyboards are not easy to use: keyboard layout doesn't much matter', *Human Factors*, **24, (5)**, 509–19.

Novak, G. (1987), 'The forgiving building: a library building consultants' symposium on the design, construction and remodelling of libraries to support a high-tech future', *Library Hi-Tech*, **5, (4)**, 77–99.

O'Grady, C. (1985), 'Australia: straining to cope with VDUs, *Computing: The Magazine*, 31 January, 16–17.

Oborne, D. J. (1982), *Ergonomics at Work*, Chichester: Wiley.

Oborne, D. J. (1985), *Computers at Work: a behavioural approach*, Chichester: Wiley.

Oslgaard, J. N. (1985), 'Automation as a socio-organizational agent of change: an evaluative literature review, *Information Technology and Libraries*, **4, (1)**, 19–28.

Osaka, N. (1985), 'The effect of VDU colour on visual fatigue in the fovea and periphery of the visual field', *Displays, Technology and Applications*, **6, (3)**, 138–40.

Pavlin, S. A., Plumanis, G. and Zvejnieks, L. R. (1983), 'Design of library facilities for the Ontario Ministry of Transportation and Communications', *Science and Technology Libraries*, **3 (4)**, 43–50.

Paznik, M. J. (1986), 'Ergonomics does pay', *Administrative Management*, **47, (8)**, 16–24.

Pearce, B. G. (1984), *Health Hazards of VDTs?*, Chichester: Wiley.

Pearce, B. (1985), 'Repetition strain injuries – the Australian experience' in *Alleged Health Hazards of Work at VDUs: current industrial relations issues. A conference held by Humane Technology in London on 26 and 27 November 1985*, Humane Technology, Quorn, 89–109.

Pearce, B. (1986a), 'Introduction' in *Job design and VDU work: a one-day conference, 1 October 1986*, Humane Technology, Quorn, 1–3.
Pearce, B. (1986b), 'Looking on the bright side', *The Guardian*, 3 April, 16.
Pearce, B. (1986c), 'VDUs – will the screen take the strain?', *Personnel Management*, **18, (7)**, 42–4.
Pearson, M. (1984), 'Why office lighting is looking up', *Business Systems and Equipment*, October, 66–71.
Peters, T. J. and Waterman, R. H. (1982), *In Search of Excellence*, New York: Harper and Row.
Peterson, L. R. and Peterson, M. J. (1959), 'Short term retention of individual verbal items', *Journal of Experimental Psychology*, **58**, 193–8.
Pheasant, S. (1986), *Bodyspace: Anthropometry, Ergonomics and Design*, London: Taylor and Francis.
Pinder, C. and Storey, C. (1986), 'Green light for new technology: the ergonomics of automation', *Library Association Record*, **88, (6)**, 282–3.
Pollatschek, M. A. and Gilad, I. (1986), 'The customizable keyboard', *Behaviour and Information Technology*, **5, (3)**, 283–7.
Project Office Furniture (1984), *Office Environments*, Haverhill: Project Office Furniture.
Quinlan, T. (1987), 'The management of change' in Work Research Unit (ed.), *Information Technology – its profitable introduction to business*, 34–41.
Ralli, T. (1988), Personal communication with the authors, University of Western Australia.
Rea, G. (1987), 'Designing an issue desk', in G. Rea and C. Storey (eds), *Ergonomics in Libraries*, Library Association, University Colleges and Research Section South Western Group, Discussion Papers no. 2, Bath.
Reeves, P. E. (1984), *The Challenge of Major Reorganization: computers in Essex County Libraries*, London: Library Association.
'Results in focus' (1988), *New Scientist*, 18 August, 21.
Rooks, D. C. (1988), *Motivating Today's Library Staff: a management guide*, Phoenix, Ariz.: Oryx Press.
Rummelhart, D. (1977), *Human Information Processing*, New York: Wiley.
Russell, N. J. (1986), *The Job Satisfaction of Non-professional Library Staff*, Leeds Polytechnic Department of Library and Information Studies Research Report no. 20. Leeds: Leeds Polytechnic.
Rycroft, R. J. G. and Calnan, C. D. (1984), 'Facial rashes among Visual Display Unit (VDU) operators' in B. Pearce (ed.), *Health Hazards of VDTs?*, Chichester: Wiley, 13–15.
Scalet, E. A. (1987), *VDT Health and Safety: issues and solutions*, Lawrence, Kansas: Ergosyst Associates.
Schneider, W. (1982), *Automatic/Control Processing Concepts and their Implications for the Training of Skills – Final Report*, Technical Report no. 8101, Champaign, Ill.: University of Illinois, Human Attention Research Laboratory.
Schneider, W. (1984), *Training High Performance Skills: fallacies and guidelines*, Technical Report no. HARL-ONR-8301, Champaign, Ill.: University of Illinois, Human Attention Research Laboratory.
Schraml, M. L. (1981), 'The psychological impact of automation on library and office workers', *Special Libraries*, **72, (2)**, 149–56.

Schrenk, L. P. (1969), 'Aiding the decision maker – a decision process model', *Ergonomics*, **12, (4)**, 543–58.

Sekey, A. and Tietz, J. (1982), 'Text display by 'saccadic scrolling', *Visible Language*, **16, (1)**, 62–77.

Sell, R. G. (1986), 'Job design – why bother? The experience of the Work Research Unit' in *Job Design and VDU Work: a one day conference, 1 October 1986*, Humane Technology, Quorn, 5–17.

Shackel, B. (1983), 'Workstation design' in Ericsson Information Systems (ed.), *Ergonomic Principles in Office Automation*, Bromma, Sweden: Ericsson Information Systems, 59–76.

Shackel, B. (ed.) (1984), *Applied Ergonomics Handbook*. Richmond: Kingprint.

Shackel, B. (1989), Personal communication with the authors.

Shaffer, L. H. (1986), 'Skilled typing performance and keyboard design', *Current Psychological Research and Reviews*, **5, (2)**, 119–29.

Shneiderman, B. (1986), *Designing the User Interface: strategies for effective human–computer interaction*, Reading, Mass.: Addison-Wesley.

Sidgreaves, I. (1985), 'Decision-making?' in J. Cowley (ed.), *The Management of Polytechnic Libraries*, Aldershot: Gower in association with COPOL, 59–83.

Slovak, A. J. M. and Trevers, C. (1988), 'Solving workplace problems associated with VDTs, *Applied Ergonomics*, **19, (2)**, 99–102.

Smith, R. (1981), *Introducing New Technology into the Office*, WRU Occasional Paper 20, London: Work Research Unit.

Somerville, A. N., Caren, L., Kaplan, I. and Reiman, D. J. (1986), 'The new science and engineering library at the University of Rochester: blending traditional with electronic', *Science and Technology Libraries*, **7, (2)**, 71–93.

Spackman, A. (1988), 'Warning on office health hazards', *The Independent*, 13 August, 4.

Sperandio, J. (1987), 'Software ergonomics of interface design', *Behaviour and Information Technology*, **6, (3)**, 271–8.

Spilling, S., Eitrheim, J. and Aarås, A. (1986), 'Cost–benefit analysis of work environment investment at STK's telephone plant at Kongsvinger' in N. Corlett, J. Wilson and I. Manenica (eds), *The Ergonomics of Working Postures: models, methods and cases. The proceedings of the first international occupation ergonomics symposium, Zadar, Yugoslavia, 15–17 April, 1985*, London: Taylor and Francis, 380–97.

Steele, C. (1986), 'New library technologies and physical problems', *Information Technology and Libraries*, **5, (3)**, 228–39.

Stewart, L. (1982), *What do UK librarians dislike about their Jobs?*, CLAIM Report no. 18, Loughborough: Centre for Library and Information Management, Loughborough University.

Stewart, T. (1980), 'Problems caused by the continuous use of visual display units, *Lighting Research and Technology*, **12, (1)**, 26–36.

Stobie, I. (1986), 'Alternative keyboards: goodbye to QWERTY?', *Practical Computing*, **9, (6)**, 52–5.

Stueart, R. D. (1984), 'Preparing libraries for change', *Library Journal*, **109, (15)**, 1724–6.

Suter, E. (1985), 'The legal aspects' in *Alleged Health Hazards of Work at VDUs: current industrial relations issues. A conference held by Humane Technology in London on 26 and 27 November 1985*, Humane Technology, Quorn, 185–95.

Sykes, P. (1986), 'Automation and non-professional staff at the Polytechnic of the South Bank, *Training and Education*, **3, (3)**, 50–6.

Tedd, L. (1980), *An Introduction to Computer-based Library Systems*, Chichester: Wiley.

Tougas, G. and Nordin, M. C. (1987), 'Seat feature recommendations for workstations, *Applied Ergonomics*, **18, (3)**, 207–10.

Turner, N. (1981), 'The attitude of non-systems staff to automated systems' in *CLAIM Report no. 8 The management of technical innovation in libraries: proceedings of a conference.*, Loughborough: Centre for Library and Information Management, Loughborough University, 13–22.

Underwood, G. (1976), *Attention and Memory*, Oxford: Pergamon.

'VDU award' (1986), *Library and Information News*, November, 6.

'VDU users – read and relax' (1986), *Mind Your Own Business*, April, 69.

Van Cott, H. P. and Warrick, M. J. (1972), 'Man as a system component' in H. P. Van Cott and R. G. Kincade (eds), *Human Engineering Guide to Equipment Design*, New York: McGraw-Hill, 17–40.

Van Nes, F. L. (1986), 'Space, colour and typography on visual display terminals', *Behaviour and Information Technology*, **5, (2)**, 99–118.

Vartabedian, A. G. (1970), 'Human factors evaluation of several cursor forms for use on alphanumeric CRT displays'. *IEEE Transactions Man-machine systems.* **mms–11, (2)**, June 1970.

Velton, E. (1968), 'A laboratory task for induction of mood states', *Behaviour Research and Therapy*, **6**, 299–342.

Waters, D. (1986), 'Assessing the impact of new technology on library employees', *Library Automated Systems Information Exchange*, **17, (1)**, 20–27.

Weise, F. O. and Borgendale, M. (1986), 'EARS: Electronic Access to Reference Service', *Bulletin of the Medical Library Association*, **74, (4)**, 300–4.

West, B. C. (1987), 'Staff management and communication', in D. Revill (ed.), *Personnel Management in Polytechnic Libraries*, Aldershot: Gower, 76-99.

Westlake, D. (1988), *Designing a Library for Information Technology*, Paper presented to the Library Association Information Technology Group annual conference, 3 July.

White, H. S. (1985), *Library Personnel Management*, White Plains, NY: Knowledge Industry Publications.

Whitford, I. (1986), 'Mouse motion', *Systems International*, September, 31–32.

Wickens, C. D. (1984), *Engineering Psychology and Human Performance*, Colombus, Ohio: Charles Merrill.

Wickens, C. D. (1987), 'Information processing, decision-making and cognition' in G. Salvendy (ed.), *Handbook of Human Factors*, New York: Wiley, 72–107.

Wickielgren, W. A. (1964), 'Size of rehearsal group in short-term memory', *Journal of Experimental Psychology*, **68**, 413–19.

Wickielgren, W. A. (1981), 'Human learning and memory', *Annual Review of Psychology*, **32**, 21–52.

Wilkins, A. J. (1985), *Visual Display Units and Fluorescent Lighting Affect Movements of the Eyes Across Text*, Report no. HF104, Winchester: IBM Hurlsey Human Factors Laboratory.

Wills, K. (1985), 'Why QWERTY is out of touch', *Computing: The Magazine*, 28 March, 9.

Yates, J. W. and Karwowski, W. (1987), 'Maximum acceptable lifting loads during seated and standing work positions', *Applied Ergonomics*, **18, (3)**, 239–243.

Zaret, M. (1984), 'Cataracts and visual display units' in B. G. Pearce (ed.), *Health Hazards of VDTs?*, Chichester: Wiley, 47–59.

Zoltan, E. and Chapanis, A. (1982), 'What do professional persons feel about computers?' *Behaviour and Information Technology*, **1, (1)**, 55–68.

Index